Soviets in Space

Soviets in Space

The People of the USSR and the Race to the Moon

Colin Turbett

First published in Great Britain in 2021 by
Pen & Sword History
An imprint of
Pen & Sword Books Ltd
Yorkshire – Philadelphia

Copyright © Colin Turbett 2021

ISBN 978 1 39900 486 2

The right of Colin Turbett to be identified as Author of this work has been asserted by him in accordance with the Copyright, Designs and Patents Act 1988.

A CIP catalogue record for this book is
available from the British Library.

All rights reserved. No part of this book may be reproduced or transmitted in any form or by any means, electronic or mechanical including photocopying, recording or by any information storage and retrieval system, without permission from the Publisher in writing.

Typeset by Mac Style
Printed and bound by CPI Group (UK) Ltd, Croydon, CR0 4YY

Pen & Sword Books Limited incorporates the imprints of Atlas, Archaeology, Aviation, Discovery, Family History, Fiction, History, Maritime, Military, Military Classics, Politics, Select, Transport, True Crime, Air World, Frontline Publishing, Leo Cooper, Remember When, Seaforth Publishing, The Praetorian Press, Wharncliffe Local History, Wharncliffe Transport, Wharncliffe True Crime and White Owl.

For a complete list of Pen & Sword titles please contact

PEN & SWORD BOOKS LIMITED
47 Church Street, Barnsley, South Yorkshire, S70 2AS, England
E-mail: enquiries@pen-and-sword.co.uk
Website: www.pen-and-sword.co.uk

Or

PEN AND SWORD BOOKS
1950 Lawrence Rd, Havertown, PA 19083, USA
E-mail: Uspen-and-sword@casematepublishers.com
Website: www.penandswordbooks.com

This book is dedicated to three 'Valentinas' who represent the best of the Soviet Union during the years of the space race:

- *Valentina Pomarova* – who should have made it into space
- *Valentina Gagarin* – the main support of the first one to do so
- *Valentina Tereshkova* – a textile worker who dreamed of driving trains but ended up doing what no woman had ever done

Contents

Acknowledgements viii
Abbreviations x
Introduction xi

Chapter 1 One for All and All for One: The Citizen in Soviet Society 1

Chapter 2 The Phoenix that Rose from the Ashes of War: The Origins of the Soviet Space Programme 21

Chapter 3 Defending the USSR: The Onset of the Cold War and the Arms Race 36

Chapter 4 First Steps to the Stars: Sputniks and Canine Scouts, 1956–1960 59

Chapter 5 Workers in Space: The First Cosmonauts, 1961–1964 73

Chapter 6 Selling the Dream 105

Chapter 7 The Race to the Moon: Failure and Retreat 132

Chapter 8 Building Communism on Earth: The Virgin Lands Campaign and the Baikal–Amur Railway 150

Chapter 9 The End of the Soviet Dream 174

Appendix
Leading Characters 194
Bibliography 200
Index 204

Acknowledgements

As with my two previous Pen & Sword-published books on the history of the USSR, the compilation of material has involved lots of collaboration and effort by others. I am again grateful to Valentina Kudinova from Kharkov, Ukraine, for her research and translations. On this occasion, she enlisted the enthusiastic help of Galina Zheleznyak, Director of the Yuri Gagarin Planetarium and Museum in Kharkov, and I am very grateful to her for photographs and information. Natalia Yurievna McAllister, a neighbour who originates from Chelyabinsk, Western Siberia, also provided very useful assistance and insight – she is also patiently teaching me Russian. Others who helped are former *Morning Star* journalist, Ken Ferguson (now editor of *Scottish Socialist Voice*), and

The author.

Natalia McAllister.

Valentina Kudinova (right) and Galina Zheleznyak (left), Kharkov Planetarium, flanked by Gagarin and Korolev.

photographer Andrea Anderson. Thanks too to my wife, Diana, and other family members for encouragement and support. Finally, the help and enthusiasm of Claire Hopkins, Chris Cocks and other Pen & Sword staff must not go unrecorded. The images and photographs are an important part of this book: permissions and licences have been obtained where necessary and credit given where authorship is known. Particular thanks to Marvel Comics for permission to use the image on page 17. In some cases, permission has been impossible to obtain and if issues arise from this, the author should be contacted via the publisher so that the matter can be properly resolved.

Abbreviations

BAM	Baikal–Amur Railway
CIA	Central Intelligence Agency
CPSU	Communist Party of the Soviet Union
DOSAAF	Volunteer Society for Cooperation with the Army, Aviation, and Navy
GIRD	Group for the Study of Reactive Motion
GNP	gross national product
HSU	Hero of the Soviet Union
IAEA	International Atomic Energy Agency
ICBM	Inter-Continental Ballistic Missile
KGB	Committee for State Security
Komsomol	All-Union Leninist Communist Youth League
MAD	Mutually Assured Destruction
MTS	Machine and Tractor Station
NASA	National Aeronautics and Space Administration
NII-88	Scientific Research Institute No. 88
NKVD	People's Commissariat for Internal Affairs
NORAD	North American Aerospace Defence Command
OKB-1	Experimental Design Bureau 1
OSOAVIAKhIM	Union of Societies of Assistance to Defence and Aviation Chemical Construction of the USSR
RABE	Missile Construction and Development in Bleichenrode, Raketenbau and Entwicklung
RNII	Jet Propulsion Research Institute
SALT	Strategic Arms Limitation Talks
SDI	Strategic Defence Initiative (Star Wars programme)
UFO	unidentified flying object
USSR	Union of Soviet Socialist Republics (CCCP in Russian)

Introduction

The 1942 Soviet film *Mashenka* is a love story that tells of a young man and woman who meet in 1939. She a telegraphist, and he a taxi driver. Alyosha, the young man, impresses Masha with his dreams for the future: he wants to be an engineer and help build the spaceships that will one day travel to other planets. She dreams of becoming a nurse and then a doctor and studies hard at night school. Their plans are interrupted by separation, and then war that engulfs both their lives. Masha, now a frontline nurse, meets the wounded hero-soldier Alyosha and the film ends with their pledge to be together again in the future, as they go off separately, to where the war needs them.

The film is direct in showing the material poverty and hardship of their lives before the war and the dangers they later face in battle. A recurring theme is space, the planets and dreams for the future. While it seems unlikely such a couple could survive the war, their fascination with a world beyond our own, shared with audiences at a time when life itself was in peril, not only characterizes the hopes and dreams of communism but also reflects the actual aspirations of some individuals of their generation who went on to realize them in the 1960s.

Fast-forward just twenty years to 12 April 1961, a date that anyone who was around during the years of

Soviet postcard with a Mayakovsky poem used to show how the space programme was inspired by the October Revolution. (*Author's collection*)

Soviet rule in Russia and its republics, and even many from later generations, still remember with undying pride. On landing from his pioneering flight into space, Pilot-Cosmonaut Gagarin stated:

> I beg to report to the Party, to the Government and to Nikita Khrushchev personally that the landing was normal, that I am feeling well and have sustained no injuries or disturbances. (Quoted in Smolders 1973, p113)

These words would become, in paraphrased form, famous as a cliché for anyone asked how their flight had been. The 12th of April later became Cosmonaut's Day – a public holiday and day of celebration. It is significant because it represents the highwater mark of what was described as 'really existing socialism' in the USSR (the popular and official description that distinguishes the present from the ideal of future communism). The story behind this achievement, at a time of intense ideological competition between the USSR and the USA, tells us much about those times and the hopes and fears that arose late in a century scarred by wars and the advancement of the machinery of death and destruction.

This is not a technical book, rather it tries to describe the social and political background to the space race and its place in the rise and fall of the Soviet Union. Its main focus is on the years of success of the USSR's efforts in the 1950s and 1960s. The history of the Soviet space programme in those years is complex and certainly does not involve a linear path of development. It involved competition and intrigue between rival designers, the favour of politicians and, at all times, secrecy. Parallel programmes ensured that some would reach fruition and success, while others would fail and suffer cancellation without details being revealed to the public. The

Gagarin as a messenger of peace. (Утро Космической Эры *1961*)

'race to the moon' was the principal focus of the space programme, both in the USSR and the USA.

The end of the Second World War in 1945 found the USSR in the position of being a reluctant superpower whose only equal was the USA. This was never a war aim. The conquest of territory and insistence on post-war influence in most of the areas of Eastern Europe it now occupied was based not on a wish to expand its borders for economic gain, but to secure them once and for all after the years of conflict that pervaded the young country's history. Its armed forces, massive though they were, had been built during wartime purely to drive back the Germans. The navy was tiny and had only played a peripheral role, and the large air force predominantly consisted of fighters and ground attack aircraft whose main purpose was to support armies on the ground. The Western Allies meantime had, many would say for no other purpose, proved their strength to the USSR through the saturation bombing of Dresden and the explosion of two atomic bombs in Japan. The necessity to do this was an implicit recognition that the main effort, to beat the Nazis and win the war, was made by the people of the USSR on an Eastern Front that had seen 95 per cent of all action in the conflict in Europe.

The fact that the USSR had by then abandoned all pretence of exporting revolution and spreading communism, was overlooked, and the interpretation of events by the Western powers saw motives in its actions that simply didn't exist. Plots for the continuation of Soviet advance westward were fabricated and quite beyond the capabilities of an exhausted though triumphant Red Army. The reality was that the Soviets had lost approximately 27 million citizens and all levels of society were sick of war and desperate for lasting peace. For Stalin, the priority was to secure the borders with the satellite countries that now fell under Soviet influence and control, acting as buffer states (de-facto agreement over which had already been made with Churchill and Roosevelt). The other principal aim was to consolidate Communist Party control internally, after the marginal relaxation of oppressive state power during the Great Patriotic War, Stalin's tried and tested methods of coercive compliance from the 1930s were brought back into play. These were grim years for the war's survivors. In contrast, the US emerged from the war having suffered no physical damage at home, a relatively small casualty rate (in comparison to the USSR and by no means diminishing the contribution and effort of American servicemen and -women) and an economy that was thriving through war production. It was of course in the interests of the businesses that had profited from the war to create in the new peace a need for continuing military development and production – hence a direct financial interest in the promotion of the Cold War.

It seems very likely that at the end of the war, the US, under a new hard-line President Truman, expected the USSR to back down in the face of the end-of-war demonstrations of military might against Germany and Japan, withdraw to their pre-war borders and allow the Western powers to dominate politics in countries the Red Army had liberated from the Nazis. However, at the Potsdam Conference in Berlin in July 1945, Stalin had stood firm and continued to do so thereafter, signalling the real start of the Cold War. The Western powers now combined militarily against their new (but erstwhile) ideological enemy and the USSR was forced to keep up and develop a similar armoury. Over the ensuing years the Soviets built up a defensive capability that tried to match the West in terms of air and naval power as well as ground forces. Its borders were made increasingly secure to keep people in and keep out Western opponents. Vast areas of the economy were devoted to military spending rather than improving the lot of the ordinary Soviet citizen. Eventually the tensions and contradictions in all this led to the USSR's demise and the apparent victory of capitalism over communism with the disintegration of the Soviet Union in 1991.

While Soviet Communism was at times utterly totalitarian, it also relied on the consent and support of the people. Efforts were continually made to keep the ordinary Soviet citizen in tune with the socialist aspirations of the state. This, of course, involved the celebration of Soviet successes, the space race with the USA, which saw fierce competition to be the first to land a human being on the moon, offered an opportunity for this. Intentions were ostensibly peaceful and made based on global human progress rather than for obvious territorial or economic gain. Naturally enough, Western reports at the time, and subsequent histories, were tinged with Cold War ideology and often missed out on essential components of the Soviet experience of the space race. This book intends to give an account of that period, particularly the years when the USSR led the way in space technology, and describe it as seen through the prism of ordinary Soviet citizens. The book is not a detailed academic history of those years (it focuses on the USSR and mentions the US only where relevant to the story) so will not reference every fact and opinion stated – it will, however, list a bibliography so the reader can turn to the sources for further enquiry. The principal source of factual information about the Soviet side of the space race comes from Siddiqi's 2000 work (generally regarded as the best and most well-researched book available), the first-hand account of Boris Chertok and the very frank diaries of Nikolai Kamanin. It also uses contemporary Soviet sources, as well as subsequent writings, that try and explain their position and experience. These include the two stories written as events unfolded by the well-informed journalists Burchett and Purdie, and the later post-Soviet memoirs of cosmonaut Alexei Leonov. The book's many contemporary images are intended to capture the feeling and meaning of the times.

The Soviet leadership had to balance the prestige gained from the highest levels of scientific achievement with difficulties supplying the basic needs of the population. All this in a centrally controlled and planned economy in which the market played little part. This was managed through careful

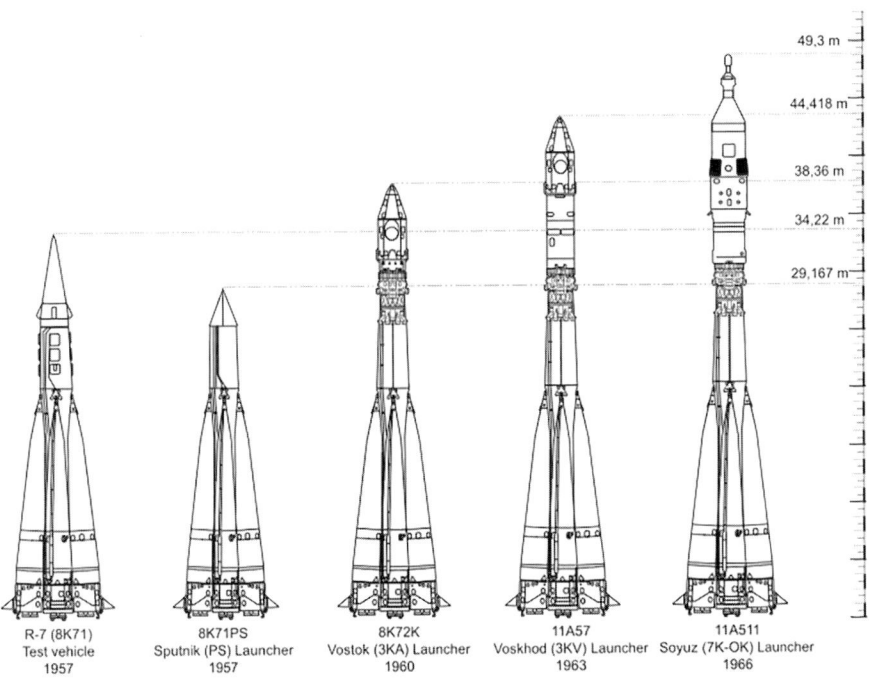

The Soviet space rockets compared. (*Nasa/Peter Gorin/Emmanuel Dissais via Wikimedia Commons*)

presentation of the facts and posing them as signals of a great future under communism. The people's active involvement was needed to realize such aspirations, just as it had been during the war when propaganda centred on the effort and sacrifice required of the whole population to drive back the invader and save the historic motherland. In the post-war era the leadership devised other plans for mass participation that would engage the population and build the economy. The principal two prestige projects that would involve the young were the Virgin Lands Campaign and the Baikal–Amur railway construction project (BAM), which are described in Chapter 8. Both are examples of physical effort to build socialism that accompanied the remote and highly specialized efforts of the space teams. Neither have any parallel in the West.

Galina Tikhomirova, Galina Obleukhova and Nina Kulikovskaya, student cooks at Rovetsnik BAM, 1982. (*V. Rodonov RIAN via Wikimedia Commons*)

A main premise of the book is that whatever were the associated military agendas, space exploration was essentially peaceful in its intentions, designed on the part of both sides of the space race to showcase their respective ideologies. For the Americans, this was to demonstrate the superiority of the American way of life with its accompanying consumerism and justify its need to dominate world markets and resources such as oil production. For the USSR this was to prove that the communist dream of peaceful progress and equality led by working people, with production for need, not profit, could result in mankind's mightiest achievements. In a speech to the 22nd Congress of the Communist Party of the Soviet Union in October 1961, Premier Khrushchev declared:

> The success of the Soviet Union and the other socialist countries have tremendous appeal. Like the rising sun, they illumine the right

Introduction xvii

road for other peoples to achieve victory for the most just social system in the shortest historical space of time. (CPSU 1961, p322)

Khrushchev went on to declare the Soviet Union's peaceful intentions:

The Soviet Union is far from seeking to dictate its will and terms to other countries. Even though we have achieved indisputable superiority in rocketry and nuclear arms, we have proposed general and complete disarmament and the destruction of nuclear weapons under the strictest international control. (Ibid., p327)

Although ignored and widely disbelieved in the West, this message would have struck a chord with a home audience for whom the memory of the devastation of the Second World War was still fresh. Achievements in the space race were things to be proud of, despite the essential fact they were being won at the cost of improving the quality of life of ordinary Soviet citizens. They still suffered a lack of choice over essential as well as consumer goods, as well as poor and overcrowded housing conditions that were only slowly improving. The necessarily secretive nature of the space race mirrored the arms race – where espionage was used by both sides to try and gain the advantage, and in the words of Lewin (2005, p385)

Images of cosmonauts at an exhibition in Poland. (*Ulrich Berchert DDR*)

'helped perpetuate the worst, most conservative features of the (Soviet) system and to reduce its ability to reform itself'. As aspirations grew while the economy declined and living standards fell, the Soviet state seemed no longer monolithic. By the late 1980s, changes that seemed inevitable, but simultaneously shocking to many, came about through reforms that opened the country up to irreversible Western influence. These led to its demise and replacement with gangster capitalism and the end of the common security enjoyed under the rule of the Communist Party. At the time, change to a market-based system was welcomed by some, but over the years many former citizens of the USSR, and not just in the Russian Federation, have come to believe that the days of Soviet achievement and success were golden ones.

The shock to the US in 1957 of the USSR's achievement of placing a satellite in orbit around the Earth – the event that marked the start of the space race – has been compared to the Chinese lead taken in recent years over the Americans with digital, especially mobile phone, technology. I think this overstates the significance of these recent events, but that previous competition eventually helped destroy the USSR. If we should learn anything as the world faces new and potentially catastrophic challenges, it is that cooperation and sharing of resources and technology will lead to better outcomes than global, profit-based, cutthroat competition. During the period described this competition was backed by the nuclear-weapon fuelled standoff between the two most powerful nations on Earth. The Cold War may be over but the stakes seem just as high today.

1980s Soviet postcard showing progress toward communism since the October Revolution. (*Author's collection*)

Chapter 1

One for All and All for One: The Citizen in Soviet Society

By working for society – you work for yourself
Labour for the Nation – the highest happiness
It is not enough to be a good performer of a given task – work creatively and truly communistically
Knowledge and experience in the introduction of modern technology is everyone's concern
A minute of work is the nation's wealth
If it's made with your own hands – it must be the best
Rejects at work are a disgrace and let down co-workers
A worker's conscience is stronger than the Quality Control
Study constantly – study is the mother of skill and with your knowledge, teach others
If you are behind, ask your friend for help – it's better to ask than fall behind
Bring beauty into labour and life
At work, at home and with friends, always remember you are a worker – do not stain your high standing
Where there are workers there is no place for hooliganism, alcoholism and parasitism

The 'Law of Honour at Work'

This chapter will start the book with a description of what shaped the Soviet people of the post-war era of space success, and what, in particular, characterized them. People are shaped by the times in which they live, the social and political arrangements of their country and their social place within it. For Soviet citizens, much more markedly than for people in the West, the way they saw themselves changed between 1950 and 1990. No one in the post-war years could have imagined that their country would become subject to the same capitalist regime experienced by people in America and elsewhere. The end of that story lies in later chapters.

The Post-War Builder of Communism

It is doubtful if anyone, other than those who experienced life under socialism in Eastern Europe and the Soviet Union, can quite understand what it all meant for the average person. Visitors to former Soviet satellite countries or republics that became independent after the dissolution of the Soviet Union in 1991 can visit museums and exhibitions that focus rather gloatingly on the worst aspects of the discarded system: the shortages, the oppression, the laughable quality of some consumer items. Some have taken it to the absurd, and dangerous, extent of characterizing fascism and communism as ideologies of similar extreme which ignores the vital Soviet role in winning the Second World War. While this features less in the Russian Federation and its friendly neighbours, positive aspects of Soviet life and culture are barely expressed and may soon be forgotten entirely as they are seen to belong to another, increasingly distant, age.

Onlookers and MiG-15 fighter planes at the Tushino airshow, August 1953. (Soviet Union *magazine, September 1953*)

One for All and All for One: The Citizen in Soviet Society

To provide a context for later chapters that describe the impact of the space programme, this one will concern itself with what it meant to be a Soviet citizen and why most of the population accepted its premise and underlying philosophies without even thinking about it, let alone bring it into question.

The average 1950s Soviet citizen had no memory of life under the Tsar before the First World War. Most had grown up in a country with a particular set of values that celebrated the needs of the collective above the individual. Competition existed and was important, but its thrust was to demonstrate how individual achievement could contribute to the greater good rather than as a means for personal advancement. Individual success was measured not by riches or personal accumulation of possessions, but by how one's work contributed to the community. Often that community was the immediate group with whom one worked – it might be a shift which undertook a particular task in a factory, or a group of workers on a collective farm, and any kind of work unit. Success for the group was measured by the extent to which their output lived up to the expectations of the plan (the production target) that had been set elsewhere. An individual contribution might result in a mention on the workplace's Board of Honour. Banners awarded to group winners of 'socialist competition' were proudly carried on annual May Day marches. Productive output was fixed according to centrally determined plans rather than the marketplace as we understand it under the Western capitalist mode of production. This led to a different mindset that created what is described here as the Soviet citizen. In the USSR the sixties generation (known as *shestidesiatniki*) characterized those born just before the war who had known hardship and loss during the war, who had seen dramatic rises in the standard of living as they entered the 1960s and were justifiably proud of Soviet achievements and unquestionably loyal to the state.

Certainly, everything in the Soviet Union was provided on an almost equal basis, the only expectation being that one worked and made a contribution – 'he who does not work does not eat'. Work was available for all and unemployment was unknown. Under-employment certainly featured and not all work was either socially necessary or especially productive. This could result in some odd manifestations of creativity. An example from the 'stagnation' period under Brezhnev is the proliferation

Soviet 'Law of Honour at Work', a banner displayed in workplaces. (*Author's collection*)

1958 Soviet postcard celebrating May Day. (*Author's collection*)

of architecturally designed and imaginative bus stops throughout the republics of the Soviet Union designed by under-utilized architects and students who had few other outlets for their talent. Waste was also an issue, as were social ills such as alcoholism and hooliganism (a very loose term in the Soviet dictionary) among the youth. Many individuals were out of tune with the system – intellectuals among them, who sought artistic and personal freedoms associated with the West. Some, like the poet Joseph Brodsky and notable scientist Zores Medvedev, ended up in psychiatric institutions that were virtual prisons (the *psikhushka*). This was not just to control such dissidents; their belief systems were seen as signs of madness. Soviet psychiatrists even came up with a diagnosis: *vyalotekushchaya shizofreniya* (sluggish schizophrenia), to explain so-called paranoid obsessions with 'truth and justice'.

Privilege existed, not for the rich as in Western society (they did not exist in the same way), but was reserved for the *nomenklatura*, a term popularized by dissident writers like Michael Voslenski. These were the officials and bureaucrats of the ruling Communist Party, for whom were reserved the best housing, *dachas* (country second homes), motor cars and

One for All and All for One: The Citizen in Soviet Society

1959 Soviet postcard celebrating new housing for families. (*Author's collection*)

other favours distributed according to rank. Social mobility depended on favour from the Party, although technical excellence was rewarded. The industrial working class was celebrated as the most important group in society, and even though the individual might aspire to a more comfortable existence than that dictated by the discipline of the factory or other industrial workplace, working-class origins (with peasant ones a close second) and experience were certainly helpful. This resulted in major differences from Western society: miners and skilled industrial workers generally earned far more than those in public, health, and education services, where salaries were fixed.

The pinnacle years of Soviet achievement were, as we shall see in the story of the space programme, the early 1960s. By that time the country had begun to recover from the war, conditions were improving and life was beginning to feel better for most people in the USSR. The 22nd Congress of the Communist Party in November 1961, celebrated all that seemed good about socialism and how superior it was when compared to the system in the West. In the unanimously agreed programme that built on previous ones, it laid out the principles that should govern the behaviour of Communist Party members as examples for the rest of society. These are worth citing at length:

> The Party holds that the *moral code of the builder of communism* should comprise the following principles:
>
> - devotion to the communist cause; love of the socialist motherland and the other socialist countries;
> - conscientious labour for the good of society – he who does not work, neither shall he eat;
> - concern on the part of everyone for the preservation and growth of public wealth;
> - a high sense of public duty; intolerance of actions harmful to the public interest;
> - collectivism and comradely mutual assistance: one for all and all for one;
> - humane relations and mutual respect between individuals – man is to man a friend, comrade and brother;
> - honesty and truthfulness, moral purity, modesty, and unpretentiousness in social and private life;
> - mutual respect in the family, and concern for the upbringing of children;
> - an uncompromising attitude to injustice, parasitism, dishonesty, careerism, and money-grubbing;
> - friendship and brotherhood among all the peoples of the USSR; intolerance of national and racial hatred;
> - and uncompromising attitude to the enemies of communism, peace and the freedom of nations;
> - fraternal solidarity with the working people of all countries, and with all peoples.
>
> (Programme of the CPSU, October 1961, pp108–9)

Communism, Khrushchev declared to congress, would be built within the next two decades:

> Communism is a highly organized society of free, socially conscious working people in which public self-government will be established,

a society in which labour for the good of society will become the prime, vital requirement of everyone, a necessity recognized by one and all, and the ability of each person will be employed to the benefit of the people. (Khrushchev, CPSU 1961, p192)

Using a space success analogy to underline the scope of Soviet achievement he described the Party's programme as being akin to a three-stage rocket:

The first stage wrested our country away from the capitalist world, the second propelled it to socialism, and the third is to place it in the orbit of communism. It is a wonderful rocket, comrades! (Khrushchev, CPSU 1961, p188)

The important point in all this was that such lofty aspirations did represent the outlook of the average Soviet citizen, even if they might not be able, or willing, to carry out all of the 'Moral Builder' principles listed. Similar incongruences exist in the West and society today: while democracies stand for freedoms most citizens support, many of us collude, in ways that are often unconscious, with the inequalities that deny these same freedoms to many citizens. The slogan 'man is to man a friend comrade and brother' was extensively displayed on Soviet workplace walls and enjoyed universal acceptance as a reminder that the individual was part of a unit and accomplishment was, above all, through the collective endeavour. Sacrifice for the common good was expected and accepted; it was part of everyone's experience over the years since the revolution of 1917, especially during the Great Patriotic War when Stalin substituted faith in the communist ideal with love for the motherland. This explains why, to the surprise of Western commentators, there was pride in a system of rule that placed investment

Khrushchev addressing the 22nd Congress of the CPSU, 1961. (The Road to Communism, *1961*)

in '*sputniks* and *luniks* rather than houses and flush toilets' (Field 1988, p132). Certainly, Soviet citizens were never able to share the consumer expectations of their Western counterparts – although they might aspire to them and be encouraged to do so. The 1957 Soviet film *To the Black Sea* mixes typically Soviet preoccupations, like ensuring the *kholkoz* (collective farm) meets its grain quota, with the dream of car ownership and holidays at the seaside, both of which were out of reach for most citizens. Car ownership, much later in the 1970s, saw 26 people per 1,000 owning cars while in the US it was 426 per 1,000 (Koenker 2013, p180).

The idea that people from particular countries enjoy commonly found national characteristics is widely accepted and often celebrated. During the war years, the national cultures and positive attributes of the Soviet people were common currency in the UK – poetry and literature were translated into English and books about the USSR and its people were widely read. In the post-war climate of the Cold War all this was turned on its head – positives became negatives to reinforce the notion that these same people were now on the other side. Anyone who visited the Soviet Union from the West during the years of the Cold War would be struck by the contrast between the commonly expressed desire for peace and friendship found among the people, and the opposite characterization of sullen aggressive intent that informed opinions back home. The USSR was never a single homogenous people – from the old Russian Empire, it inherited vast territories inhabited by an estimated 140 nationalities. Historically dominant were the ethnic Slavs of European Russia, who were closely related to other Slavic nationalities such as the Ukrainians and Byelorussians – together they vastly outnumbered all other ethnic groups. Efforts were made to ensure minority representation in the Soviet constitution but nationalist resentment at Russian dominance was a major factor that contributed to the breakup of the USSR in the 1989/90 period.

In the 1950s and 1960s American sociologists and anthropologists published studies that tried to explain the loyalty of Soviet citizens to their regime (see, for instance, Kluckhohn 1961) so that they could understand their adversaries, and from a military and propaganda point of view, exploit their alleged weaknesses. These came up with findings that put a pejorative twist on what might otherwise have been regarded as assets: Soviet people needed a strong leader, were essentially subservient, less concerned about personal achievement and how others saw them, and more worried about the collective good and such values as loyalty and honour. This is in contrast to the image of the typical American citizen who places self and family above other considerations. A British sociologist (MacCrae 1961) comments that 'The charitable man can find inspiration in Marxism only when he is preternaturally gullible. And since this charity is often as innocent and genial and straightforward as it can be, it is not enough to recognize in Marxism only the perverted love which conceals itself in cruelty'. From such 'science' the Cold War was perpetuated in a haze of rhetoric that characterized the Soviets as enemies because their ideology and belief systems were said to be based on different and less human values than those of the average US and British citizen.

Women in Soviet Society

To be female in the Soviet Union involved mixed blessings. The revolution of 1917 included among its Bolshevik leaders, feminists of the calibre of Alexandra Kollontai who soon ensured that equality of the sexes was an important agenda item for the new regime. This was to involve social legislation that would free women to enjoy the same rights as men, including control of their bodies and freedom to engage and disengage in marriage and partnerships. It would also offer social provisions such as child care and health services that would remove the burdens associated with traditional family life. By the mid-1930s, much of this had been jettisoned as the USSR fought to catch up with the industrially advanced nations of the world: women were needed in the workplace but also in the home to take care of children and worker-husbands. Wartime gave something of a boost to women: unlike the other belligerents, many in the USSR took up arms and fought on the frontline. However, this was short lived as their reproductive role was needed after the war to increase the population and replace the generation lost to the conflict. Even so, equality in the workplace continued and women enjoyed access to a range of work roles that were denied to most in the West. Just as the character of women in Western capitalist society is reflected and defined for many by popular women's magazines in terms of fashion, identity and interests, so it was in the Soviet Union. The state-controlled magazines *Woman Worker*, *Woman Peasant* and *Soviet Woman* performed this role. Where the purpose of Western magazines was to shape women as consumers, the Soviet ones were designed to promote

10 Soviets in Space

Fanna Barmaleeva hydroelectric power station truck driver, 1962. (Za Rulem *magazine*)

Red October chocolate factory Moscow, 1961. (Soviet Union *magazine, 1961*)

acceptance that they were both workers and homemakers, and to build and celebrate the skills they required for such essential tasks (Ryabakova 2017). Some have argued that the freedom from economic dependency on men, however, limited in terms of wider freedoms, was such that women and men in the socialist countries enjoyed more equal relationships than typically found in the capitalist ones. This led to more satisfactory sexual relationships (see Ghodsee 2018 for a well-argued and grounded case for such a view). Whether all women wanted equality is a question raised by women from the Soviet era – many preferred men to take the lead in relationships (see the discussion about the popular movie *Moscow Does Not Believe in Tears* below). Another undoubted reality is that men did not change much under socialism and typically expected women to bear the brunt of family tasks.

Humour

Many aspects of Soviet life were affected by censorship – indeed the secrecy surrounding the space programme engendered styles and forms of public projection that will be examined in Chapter 6. This pervaded many aspects of life and although perhaps never reaching the levels of

One for All and All for One: The Citizen in Soviet Society

paranoia and surveillance that it did in East Germany, where one in four citizens was said to be Stasi (secret police) informers by the 1980s, Soviet citizens were aware that criticism of the state was not tolerated publicly and could have consequences. One way of dealing with this was through humour. Funny stories took on a life of their own and were passed from mouth to mouth – a fairly harmless outlet for feeling and expression. These changed in style through the years of Stalin and then the various regimes up until the final periods of stagnation and *glasnost*. Some poked fun at the ruling party and its politicians, others reflected the shortages of goods and realities of everyday life. Some ridiculed censorship itself:

> A frightened man came to the KGB. 'My talking parrot has disappeared.' 'That's not the kind of case we handle. Go to the criminal police.' 'Excuse me, of course I know that I must go to them. I am here just to tell you officially that I disagree with the parrot.'

Another genre of joke purports to be quotes of broadcasts from Radio Yerevan in Armenia – the subtext being a pejorative view of this nationality shared among European Russians, the dominant group in Soviet society. These jokes typically parody a programme where questions are sent in by listeners and answers read out in a deadpan voice.

> 'Dear Radio Yerevan, I don't know what's the matter with me. I don't love the Party any more. I feel nothing at all for Comrade Brezhnev or any other leaders. What should I do?'
>
> Radio Yerevan answered: 'Please send us your name and address.'
>
> 'Dear Radio Yerevan, is there a censorship of press and radio in the Soviet Union?'
>
> Radio Yerevan answered: 'In principle, no, but it is unfortunately not possible to go into this question in any detail at present.'
>
> 'Dear Radio Yerevan, what is an exchange of opinions?'
>
> Radio Yerevan answered: 'When you walk into your boss's office with your opinion and walk out with his.'
>
> (Quoted from *100 Best Russian Jokes*: www.johndclare.net/Russ12_Jokes.htm)

12 Soviets in Space

The Soviet Citizen in Popular Culture

One of the best and easiest ways for contemporary observers to understand how it felt to be a citizen in the post-war USSR is to look at portrayals of life in the movies and popular literature. The examples chosen for examination here are among those that were popular because within them people could recognize both themselves and characteristics of the world in which they lived.

The movies were generally light-hearted and helped people laugh at themselves in a manner not untypical of self-deprecating working-class humour internationally. The three selected here are still very popular in Russia, particularly among the generation that looks back on Soviet times with varying amounts of nostalgia. Each recounts stories about the lives of ordinary people of the *shestidesiatniki* generation.

Moscow Does Not Believe in Tears is a film made in 1980 which tells the story of three women over twenty years from the late 1950s to the late 1970s. It starts with them aged about 20, sharing a room in a hostel in Moscow. All are from the country, working in ordinary jobs but pleased to be in the big city where horizons might be broadened. One of the girls, Antonina, is already in a steady relationship with a young man from a similar background to herself and the two soon get married and have children – their lives remain steady and unremarkable; they are content and happy. Katerina works as a machine operator in a large factory, having failed her first-year university exams, but continues her studies in her spare time so that she might ultimately make more of her life. The third friend, Ludmilla, who works in a bakery, has a different, comically portrayed, obsessive ambition to put her country origins behind her by marrying a successful man. We are given insights into Soviet life and culture – the fun of communal life in the hostel, and leisure activities including a visit to the workers' *dacha* being built by Antonina's parents outside the city. The early part of the story revolves around the flirtatious (but sexually careful) Ludmilla's attempts to woo men she regards as eligible bachelors by pretending she and Katerina are the daughters of a university professor. These are, respectively, a poet (with unsympathetically portrayed dissident attitudes), a famous hockey player, and a TV cameraman. The result of this is that Ludmilla eventually marries the hockey player, who becomes an alcoholic, and Katerina embarks on a brief relationship with the TV cameraman, which despite her better judgement, is based on the premise that she is from a more cultured social background than she is. By the time he finds this is untrue and ends their relationship, she is pregnant. We see her in her factory, being spotlighted for her initiative and example to other workers. This, and the emblazoned slogans, that surround the shop floor would have been familiar to Soviet viewers. We fast forward twenty years. Ludmilla, as flirtatious as ever, is divorced from her broken alcoholic husband. Antonina is still happily married. Katerina, having refused the abortion suggested by the appalled socially superior mother of her lover, has gone on to successfully raise her child as a single parent. She has also worked hard, despite all obstacles, and is now a senior manager and important local Communist Party official. In her private life, she has an ongoing affair with a married senior official but ends it when he makes clear his choice not to disrupt his family life. After an unsatisfactory consultation with the manager of a dating bureau, where she pours scorn on the general failings of men in the USSR, Katia meets a new romantic interest – a skilled worker from whom she hides her senior position. We find out why when he discovers her status and the significant fact that she earns more than him; he leaves her as he finds this role reversal unacceptable. The film ends with their reconciliation and Katerina's capitulation to the terms he sets out: he has to be the alpha male and prime earner if their relationship is to continue. This is rather surprising to a modern Western viewer in step with notions of sexual equality, who would instinctively have admired Katerina's efforts and achievements and be unimpressed by her would-be suitor. However, to Soviet audiences, this was exactly what they wanted to see, and the male actor concerned became an overnight star and the dream of many Soviet women. In the USSR equality existed in the workplace but was not expected or even desired within a marriage. The reader might remind themselves of this when, in Chapter 5, the story of Valentina Tereshkova, the first woman in space, is related. The film was not endorsed by the Communist Party until it became popular, but it nonetheless contains cultural digs at the West. *Moscow Does Not Believe in Tears* won an Oscar award and was considered in the West to offer insight into Soviet life and values. US President Ronald Reagan, an ex-Hollywood movie actor who set great store on such things, watched it before meeting Mikhail Gorbachev of the USSR for the first time in 1985.

Postcard for the film *Moscow Does Not Believe in Tears*, 1980. (*Author's collection*)

The second of our movies is a romantic comedy, *Office Romance*, from 1977. This features various white-collar workers in a large open-plan government office whose function is to compile statistics. The sheer boredom and pointlessness of such work is parodied – comedic to both Western and contemporary Russian audiences, but familiar to the Soviet viewer in what was now the period of stagnation. The comedy revolves around the efforts of a very ordinary and rather hapless clerk, Anatoly, to woo the manager of the office, Ludmilla – a plain and hard-working woman who initially eschews emotional reward in favour of work. Against a background of a bustling but rather depressing and rainy autumnal Moscow, we meet other characters whose lives intertwine with the office. The movie director introduces numerous small touches that highlight social aspects of this period – the sheer ordinariness of people's lives and the efforts made to bring colour and hope into existence. Anatoly is described at one point by Ludmilla as a hooligan – he is anything but. This epithet in Soviet life was used to describe almost anyone whose behaviour went outside the norm. Then there is the rather slick assistant manager who has worked in Switzerland and impresses his colleagues with items he has brought back, including foodstuffs and cigarettes. At one point he takes a pair of windscreen wipers out of his bag to attach to his smart and desirable Volga car before he drives off in the rain. The trade union activist is a nosey and gossipy woman whose office welfare efforts are grudgingly tolerated by staff. The black market in Western goods and the queues to buy food are all portrayed. The hero of the movie is Anatoly, the bumbling single-parent clerk. Soviet audiences loved his character, which crops up in many Russian films and books, from all and not just Soviet eras. His ordinariness and victimhood reflected their own lives when the heroes the state might want them to emulate seemed remote. The contemporary viewer might be struck by the social solidarity shown in the movie; although the managers enjoy noticeably better lifestyles than the majority of staff, they are accessible, respectful to everyone and not socially distant.

The third film in our short study is *The Irony of Fate (How Was Your Bath?)*, a romantic comedy made in 1976 and so popular that it is still shown on Russian TV every New Year. It concerns the New Year holiday meeting of a couple in their thirties: the man, a surgeon called Zhenya, and the woman, Nadya, a teacher. The plot revolves around their chance meeting when a very drunk Zhenya is mistakenly bundled by his similarly drunken friends onto a flight from his native Moscow to Leningrad on New Year's Eve. After landing and not realizing he is in a different city, he orders a taxi to his home and is taken to an identical address in Leningrad. He goes up in the lift to the apartment and lets himself in with his key and then falls asleep. The apartment is occupied by Nadya and (we later learn) her mother. Her plans to spend the New Year with her jealous boyfriend, Ippolit, are interrupted by the uninvited Zhenya, who had similar plans to spend New Year with his fiancée Galya in Moscow. Humour surrounds the similarity of their apartments – not just the addresses and style of these *Khushchyovka* flats, but the décor and furnishing and the type of lives these two single people in their mid-thirties lead (an age when most would expect to be married). Again, a Soviet audience would have taken for granted the oddities for a contemporary audience – the scarcity of housing for single people so family groups stayed together; the nostalgia for communal bathhouses despite the novelty of now having one's bathroom; the sameness of accommodation (344 square feet) and available furnishing; the references to low salaries for these professional workers (notably Zhenya whose lifestyle in the West, even in the 1970s, would have been very comfortable by comparison). There is even a reference to the sameness of cinemas and movies at the start of the film that would have amused. Nadya is said by her friends to be a well-thought-of teacher with her name on the Board of Honour – a sought-after accolade.

All these movies had to be approved of by the authorities – they had to be devoid of criticism of the state, but satire was possible. What Soviet audiences did not know at the time was that because Nadya was played by a Polish actress (Barbara Brylska) who had a slight non-Russian accent, her voice was dubbed with the voice of a Russian actress – Russian national chauvinism was alive and well in 1977. This was not just reflected in popular culture, but in a very real sense with the dominance of Russian (and to a lesser extent Ukrainian and Byelorussian) men in positions of power throughout the Soviet Union.

A book that was immensely popular in Soviet times, although little known outside the USSR, was *Two Captains* by Veniamin Kaverin (1902–1989), first published in 1944 and reprinted forty-two times over the next twenty-five years. It concerns a long-lost polar explorer, Captain Tatarinov, and the intertwining of his life with that of Sanya Grigoriev, who is first introduced as a child. Sanya encounters poverty and hardship before fulfilling his ambition of becoming a pilot in the 1930s and towards the end of the book he is a wartime air force hero. The theme of discovery and achievement in the face of odds – both human and physical – echoes both the Soviet ideal and that of a later generation who made the first space explorations. Sanya, like his hero Tatarinov, fills his life with hard work, self-education and love of country. Naturally, for a Soviet-era book, the bourgeoisie

Postcard from film *The Irony of Fate: How was Your Bath?*, 1976. (*Author's collection*)

One for All and All for One: The Citizen in Soviet Society

are regarded suspiciously and the qualities of honest working people are celebrated, but this is also a book about love, loyalty and perseverance. Although reflecting Soviet patriarchy, the book has some strong female characters. *Two Captains* was written over five years, and its later descriptions of wartime life give a real sense of the trials, sorrows and pride of the Soviet people at that time.

Heroes and Ordinary Citizens

In Western societies, heroes (I use the word to describe heroines too) are diverse entities and do not necessarily enjoy a popularly held understanding and appreciation of their heroic status. To one individual a famous footballer might be a hero; to another, a singer or band member enjoying cult status. In the Soviet Union heroes were cultivated by the state to exemplify virtues that it was hoped all citizens might aspire to emulate. The popular fictional superheroes (and supervillains) of 1940s and 1950s American comic books (Batman, Superman, Superwoman, etc) were created to spearhead the fight of good against evil – personified during the war years by the Nazis. The evolution of several (notably Captain America) into anti-communist crusaders after the war represented the capitalist antithesis to the ideal communist hero of the Soviet years. Such heroes embodied All-American values and defeated evil enemies (not just Soviet ones) who in the popular American imagination, wanted to take over the world. These imaginary heroes were crude and comical but inspired American children and, through their movie descendants, are still with us today.

'Captain America commie smasher', US superhero, May 1954. (Captain America © 2021 Marvel)

Soviet heroes were real people even if their stories were sometimes embellished. Starting with those who performed outstanding feats during the Civil War following the revolution of 1917, the genre moved on in the 1930s to highlight those 'shock workers' like miner Alexey Stakhanov, who broke all production records to produce a record output of coal in one highly publicized shift in 1935. It also included pioneering aviators and Arctic explorers. A few years later the wartime fight against the Nazi invaders produced thousands of heroes, many recognized posthumously. In the period covered by this book, as we shall see in later chapters, the domestic market for heroes was cornered by cosmonauts. What all these heroes shared, from 1934 until its abolition in 1991, was official Hero of the Soviet Union (HSU) status. There is no equivalent in Western societies, although systems of military and bravery awards, and civic honours, certainly exist in various forms such as titles bestowed by the British monarch for particular contributions to society. HSU status involved a medal that represented the highest award of the state, a pension and other benefits, and importantly, celebration involving the telling of their stories throughout the country. By the end, HSU status had been awarded 12,777 times, of which 11,635 were awarded in the Great Patriotic War. This was a harder feat for women, only ninety-five of whom were ever awarded HSU status – forty-nine posthumously during the war. Heroics (with exceptions) were a male preserve! Among the forty-four foreign recipients was Ramon Mercader, the 1940 assassin of Stalin's arch-enemy Trotsky in Mexico. He was given the award upon his release from prison and move to the USSR in 1961. In among the real heroes from the ranks of ordinary people were, in later years, time-serving political leaders like Brezhnev who awarded himself this status on several birthdays. That last fact notwithstanding, the vast majority of HSU awardees were ordinary citizens who had made sacrifices, many involving the giving of their lives, because of their love of the motherland, the state, and its aspirations, including the common good and the international betterment of humankind. All Soviet cosmonauts received this award on each occasion they went into space.

Children were fed on the stories of HSU awardees (some of whom were also very young) from an early age in the USSR's official uniformed extra-curricular organizations like the Young Pioneers who, in 1974, had a membership of 25 million children aged between 9 and 15 (younger children could join the Young Octobrists). Their activities involved the promotion of communist ideals and atheism as well as the cultural,

One for All and All for One: The Citizen in Soviet Society

Soviet hero Gagarin with admirers at the 22nd Congress of the CPSU, 1961. (Утро Космической Эры, *1961*)

sporting and fun activities many of us would associate with the Scouts and Guides movement in the West. At the age of 15, young people could join the Komsomol, which could in some cases, pave the way for Communist Party membership for the over 28 age group – a selective mass organization for which membership had to be earned through work and example as well as political correctness. The Young Pioneers, in particular, promoted the Soviet state's official heroes as exemplars for children to study and emulate. Until the advent of the cosmonaut with Yuri Gagarin in 1961, most of these were from the Great Patriotic War – partisans, soldiers, sailors and pilots. As most were men, the women, being fewer in number, were perhaps less obscure and more celebrated among girls. Of note were the women pilots of the all-women 46th Guards Night Bomber Regiment who flew primitive U-2 biplanes, twenty-three of whom were HSUs; and the partisan Zoya Kosmodemyanskaya, an 18-year-old murdered by the Nazis early in the war. While wartime women fighters were deliberately erased from Soviet history in the early post-war years, some, like Zoya, were studied and their stories told over and over again. This ambivalence about women was part of the Soviet experience – equality of opportunity in the workplace might exist, and even enjoy guarantee in the constitution, but heroic endeavour and opportunity were mainly male preserves. This was reflected in the Soviet space programme.

Achievement in sport also held an important place in Soviet society. Sporting success extolled the health in mind and body that the socialist state hoped to achieve for everyone as the dawn of communism approached. It also encouraged the notion of socialist competition – the idea that success should be based not on selfish self-gain but in a comradely fashion through hard work and effort. Of course, these ideals might not be found in practice where winning meant everything (especially at a prestigious international level) but self-sacrifice for future benefit was encouraged through the youth organizations where sport and healthy activity (including some like parachuting and skydiving that had military connotations) were promoted for the masses. Of importance in this connection was DOSAAF (Volunteer Society for Cooperation with the Army, Aviation and Navy), which was linked with the Komsomol and provided opportunities eagerly seized by young working people like future cosmonaut Valentina Tereshkova.

* * *

As we shall see in the final chapter of this book, the scientific achievements and improved social conditions that brought hope and pride to people's lives became jaded by the end of the 1960s. A new generation was emerging in the period of stagnation (outlined in Chapter 9) during the Brezhnev years. Young people became increasingly cynical about the lack of consumer items when compared with the West, tired of stories about the war from previous generations, and felt little loyalty to the USSR as an entity or ideology. Such ideas and thoughts were accentuated among national and ethnic minorities. The brief period of pride in Soviet achievement was over.

Postcard of Soviet man conquering space. (*Author's collection*)

Chapter 2

The Phoenix that Rose from the Ashes of War: The Origins of the Soviet Space Programme

Human society must develop harmoniously. Man will continue to unlock the secrets of space and utilize its incalculable energies. That will multiply his own powers. To be able to use them rationally he must strive towards a new, higher level in his intellectual development.

The Land of Soviets is proud in the knowledge that its sons and daughters have blazed the trail into space and made a fundamental contribution to the exploration of the Universe with the aid of rocket-propelled vehicles.

V. Glushko, 1973

Konstantin Tsiolkovsky: The Father of Soviet Rocketry

Technological advance in human history is often, and sadly, driven by warfare. The Second World War certainly proves this dictum. Progress (or arguably, regress) was made in methods of human destruction that involved not just technical innovation – exemplified in the extreme form with the atomic bomb developed by the US, Britain and Canada – but also the simple barbarism of the Holocaust. The German experience under the Nazis took in two extremes: racist ideology and the use of genocidal methods of murder reminiscent of previous civilizations from history, and a futuristic outlook on the development of rocketry and space technology created with the merciless use of slave labour.

During the war the Allies largely ignored rocket-science development as a method of delivering warheads long distances, in favour of mass heavy bomber fleets. These involved larger and ever-more devastating bombs that could be released from the air by increasingly bigger aircraft, culminating in the horror of *Fat Man* and *Little Boy*, the atomic bombs dropped on Japan, the consequences of which are explored in the next chapter. The Germans, however, invested heavily in ballistic missile development and succeeded in producing and using over 3,000 V-2 (or A-4) missiles against Allied targets in the final nine months of the war. These were fortunately too late to affect the war's outcome but represented a terrifying ordeal for those upon whom they fell from high in the sky with no effective warning. The Soviets meantime, as explained briefly in the introduction, focused on ground forces with associated air power to support their drive into Germany to bring the war to an end. After the conflict, when inspecting the heavily bombed sites in eastern Germany where the Nazis had developed the A-4 missiles, the Soviets found a copy of a book previously owned by Wernher von Braun, one of the programme's chief engineering scientists. It was by the Russian, Tsiolkovsky, and had been written at the beginning of the century outlining his ideas about the future possibilities of rocket technology, including manned space flight and interplanetary travel. Von Braun's copy was laced with his notes and comments.

Konstantin Tsiolkovsky (1857–1935) was an eccentric visionary, early rocket scientist, philosopher and career schoolteacher. Although a supporter of the Soviet regime in very general terms, his ideas about eugenics (shared, with appalling consequences, by Hitler and the Nazis) and almost religious belief in the power of cosmic forces, had him regarded ambiguously in Stalin's regime of the 1930s. Knowledge of his work had him elevated to the Soviet Academy of Sciences soon after the Bolshevik Revolution. Tsiolkovsky never built a rocket, but his ideas had a marked influence on a younger generation of Soviet technicians and scientists, including Sergei Korolev and others who feature in this story. He continued to develop his thoughts about space travel throughout his life. His propositions were remarkably forward thinking and his predictions and theories about how problems could be overcome were eventually proven in practice. His early ideas about the design of spacecraft bear a resemblance to those that were later developed by the Soviet space team. In the years of Soviet space-race success, Tsiolkovsky was to be venerated as the father of Soviet space travel, although this hardly represents reality as his interactions with his young protégés have been exaggerated in that narrative. His ideas, during his lifetime in the Soviet era, were featured in exhibitions, and his models and drawings would certainly have made an impact on the young people who would later lead the Soviet space programme. In 1935, a few months before his death, a speech by Tsiolkovsky was broadcast from the podium at the Moscow May Day Parade in the presence of Stalin – a fact that received much attention in later years, as did the state funeral he was given. During the earliest years of the Soviet regime, idealism about the communist project was encouraged with the popular and artistic expression of scientific ideas about the future. Inevitably, most of this would be overshadowed during the 1930s by the pragmatic emphasis placed on Five Year Plans with the struggle to develop industrial capacity and feed the burgeoning cities through the collectivization of agriculture. Dreams of the future were permitted – but the emphasis was very much on the present with all its problems.

Konstantin Tsiolkovsky. (*Wikimedia Commons*)

In the 1930s, with the young Soviet regime making strides in aviation, particularly long-distance record setting, designers were principally focused on aircraft design, but with some interest in jet propulsion, rocketry and exploring the upper atmosphere of the Earth. Among the team so involved were the young scientists and engineers Sergei Korolev, Valentin Glushko and Mikhail Tikhonravov – all employed by the Group for

Tsiolkovsky's design for a rocket ship, 1903. (*Novosti* Soviet Man in Space, *1961*)

the Study of Reactive Motion (GIRD) established in 1931 in Leningrad. This was combined with another organization to become the Research Institute for Rocket Propulsion (RNII), whose main contribution by the outbreak of war was the Katyusha missile, used extensively and effectively by the army as a simple truck-launched attack system: large numbers of crudely targeted explosives being rained simultaneously on areas under assault. Another talented member of the team was Friedrich Zander, who was keen on space-rocket development and specialized in jet propulsion, but he died of typhus at the age of 46 in 1933.

Mikhail Tikhonrarov, 1925. (*Wikimedia Commons*)

Sergei Korolev and His Associates

As a child and young man in Ukraine, Sergei Korolev (1907–1966) developed an interest in aeronautics and gliding, and his vocational education was accelerated by the Soviet state's need for engineers and scientists to achieve its economic and industrial expansion plans. After graduation in aeronautical engineering, Korolev went to work in the aircraft design bureau led by Andrei Tupolev, and from there in 1931, to GIRD, which he was appointed to lead in 1932. The amalgamation that created RNII in 1933 brought Korolev into close contact with space enthusiast Valentin Glushko. Korolev became chief engineer at the Institute and was involved in jet and rocket propulsion development. His principal skills though were in managing and marshalling the input of others to successfully develop programmes. In 1938, at the height of Stalin's purges, his career and that of Glushko came to an abrupt halt when they and most of the leadership of the Institute were arrested on suspicion of 'wrecking' activities. It was later said that all were denounced by their colleague Andrei Kostikov, who went on to become the new head – until he was later accused of financial irregularities and removed. However, it also seems that most involved, probably including Glushko and Korolev,

The Phoenix that Rose from the Ashes of War 25

GIRD scientists, 1931. Korolev is seated at centre. (*Wikimedia Commons*)

were forced through torture to denounce one another. The Soviet writer Roy Medvedev speculated (1976) that Korolev fell out of favour because he preferred rocket design work to the aircraft development expected of him by the state – irony indeed considering the later impact of his efforts on Soviet prestige and influence. Several of those arrested, including Institute heads Kleymenov and Langemak, were executed. In a story not untypical of those terrible times, this set back the work of the RNII, and thus the interests of the USSR, considerably, and enabled the Germans to take a lead in rocketry and jet propulsion.

Korolev was tried and sentenced to ten years in late 1938, by which time the purges were subsiding after the removal (and execution) of NKVD chief Yezhov. Korolev was sent to the Kolyma goldmines in north-eastern Siberia; since his arrest, he had retained faith in Stalin and the system and had repeatedly appealed his innocence to the authorities. His experiences seem not to have dented this faith. Kolyma truly was a terrible place in one of the most inhospitable locations on Earth, where temperatures plummeted to −45°C and more in winter, and which took many months to reach by train, boat and forced march – a journey that ended in death for the weakest before life in the camp had even begun. The colony was established after the discovery of gold deposits in the 1920s, useful for the Soviet state's need to purchase goods from abroad required for its rapid industrialization plans. The Gulag camps were set up to provide slave labour for this state enterprise and existed from the early 1930s until the 1950s, by which time untold hundreds of thousands of prisoners, many of whom were later declared entirely innocent, had perished. Korolev was said to have been subject to extreme privation in Kolyma – he lost all his teeth through malnutrition, suffered scurvy and a broken jaw that was in evidence throughout his life. Within a few months he was in an extremely weakened state and would, like a third of other inmates, have died, had his talents not been belatedly recognized as necessary for the coming war against the Nazis.

Korolev was retried on the orders of the new NKVD head, Beria, in 1940, and his sentence was reduced to eight years. After a perilous six-month journey back to Moscow he was sent to a *sharashka* – a special prison for intellectuals and scientists – where he was put to work in an experimental design centre under the direction of a fellow prisoner, aircraft designer Andrei Tupolev, with whom he had worked in the early 1930s. Valentin Glushko was also imprisoned in a similar facility and the two were reunited in Kazan in 1942, where they worked together

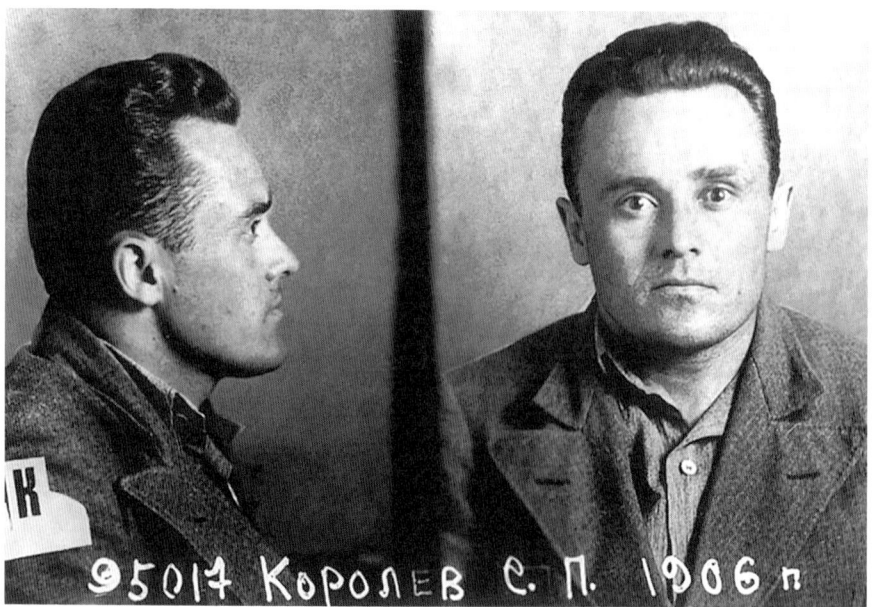

NKVD photograph of Korolev after his arrest in 1938. (*Wikimedia Commons*)

The Phoenix that Rose from the Ashes of War

A Kolyma mine in the 1930s. (*Wikimedia Commons*)

on rocket propulsion to assist propeller-powered aircraft and weaponry. Release for Korolev, along with Glushko, Tupolev and others, came in June 1944, although he remained technically captive until charges were dropped in 1957. All of this was not known to people in the Soviet Union until the late 1980s and *Glasnost*; one of the first Soviet accounts of the space programme (Riabchikov 1971) that offered details of Korolev's life and contribution, omitted any reference to the years 1938–1941.

As we shall see, Korolev was soon involved in the retrieval and appraisal of German rocket material recovered from their production sites after war's end. Korolev's organizational abilities propelled him into a leading role as chief designer for the Institute charged with missile development, and ultimately the space programme. His effective rehabilitation was completed when, in 1953, he was admitted to Communist Party membership – an unusual honour for a former prisoner. However, if this had not happened, it is unlikely he could have continued as chief designer. Korolev remained in this increasingly important role until his sudden death at the height of his career in 1966. His identity was kept secret, even from many involved in the space programme, until after his death – ostensibly to avoid assassination at the hands of the Americans. Korolev is now publicly acknowledged and celebrated as the guiding force and principal individual responsible for the years of Soviet space-race success. His talent lay not in the actual designs – he had a team of the best around for this purpose – but in the vision, passion and organizational skills required to argue a case and then bring together the expertise and resources needed to see a project through to conclusion.

Several of Korolev's associates from his early years in the Tupolev bureau feature throughout this story. Valentin Glushko (1906–1989) from Ukraine, inspired by Tsiolkovsky, with whom he corresponded in 1921 at the age of 15, was interested in space flight from an early age. He wrote his first published article on moon exploration in 1924, and two years later had a more detailed article entitled *Extra-Terrestial Station* published by a science and technology magazine. Glushko was said to have demonstrated exceptional technical abilities when he went to

Sergei Korolev, 1965. (*Wikimedia Commons*)

Katyusha rockets. (*Wikimedia Commons*)

work at GIRD, having failed to graduate from university. He too was arrested in 1938 but seems to have fared better than Korolev (some say because he denounced his comrades). He was transferred, after conviction and an eight-year sentence in 1939, straight to a *sharashka* design bureau, where, as chief of a department investigating rocket engines, he was reunited with Korolev. Another notable rocket engineer of the period associated with the group was the Russian Mikhail Tikhonravov (1900–1974). His career in GIRD and then RNII continued through the 1940s until he became an important part of Korolev's team in the 1950s. He was particularly associated with the Katyusha weapon system that was deployed so successfully during the war. Other pioneering Soviet rocket scientists from this early period died in the war along with 27 million of their compatriots. These included Kondratuk (killed in a battle near Moscow), Kisenko, Perelman (who died during the long siege of Leningrad) and Rynin.

Valentin Glushko. (Development of Rocketry & Space Technology in the USSR, *1973*)

A prominent member of the team assembled soon after the war was Boris Chertok (1912–2011), an aeronautical engineer since the early 1930s, who became head of the control systems section and deputy to Korolev in 1956. Chertok would write an important insider history of the Soviet space-race effort.

The Race for Nazi Rocket Technology

At the end of the Second World War, space travel was far from anyone's mind. From amidst the ruins of Germany, the Allies were set on ensuring that fascism was properly defeated by destroying and removing all vestiges of offensive military hardware and gathering secrets and technology they could for themselves before others got to them first. The Soviets also wanted to salve what they could as reparation for the wanton devastation wreaked upon their country by the Nazi invasion and subsequent battles.

This ranged from official decrees which stipulated limits on personal booty for troops of different ranks, to the dismantling of whole factories and their removal to the USSR. The development sites and manufacturing facilities for the German missile systems were all situated in the east of the country – as far as possible from the UK-based heavy bomber fleets whose task, among others, was to destroy the Nazi armament industry. This placed most within Soviet-occupied territory.

The original development site for the German missiles was at Peenemünde on the Baltic coast. It was here that the world's first rockets with space potential were built to carry warheads. The site became known to the Allies and was extensively bombed in August 1943 and July and August 1944. By mid-1944 the production facility had transferred to a mountain site in Mittelwerk, Bavaria, which had its own concentration camp, Dora, from which slave labour was supplied to build a vast complex of underground tunnels, hidden from the air and unknown to the Allies. Mittelwerk was in a remote part of a region captured by the US Army and scheduled to be handed over to the Soviets as per the immediate post-war agreement over zones of occupation. Its discovery, on 10 April 1945, was by a US Army private who had followed a railway line for over 100 miles on foot to what he believed would be a concentration camp. His suspicions were confirmed by the grim discovery of Dora, and the decaying corpses of many of its inmates, as well as the nearby entrance to the tunnel system they had died constructing. Knowing that the site would soon be in Soviet hands, the US Army quickly moved in to remove everything of technical interest. Brzezinski (2007) writes that this amounted to 16 cargo ships' worth of equipment and parts, including 100 intact V-2 (or A-4) rockets that were all taken to New Mexico. The US Army even removed the lights so that when the Red Army arrived on 14 July 1945, they had to contend with darkness to look into what was now an empty site. However, when the Soviet scientists arrived, including Boris Chertok, they were met with survivors of the mainly Soviet POW slave labourers who had built the complex. They were led to parts that had been successfully hidden from the Americans, including some important and highly innovative components such as a gyro-stabilized platform.

The US Army, however, in the long-running Operation Paperclip, had made off with the principal engineering architects of the German rocket programme, including Wernher von Braun, the 33-year-old chief

The Phoenix that Rose from the Ashes of War 31

American soldiers at Mittelbau/Dora, July 1945. (*US Army via Wikimedia Commons*)

engineering scientist. As an SS officer who had worked within the complex where thousands of concentration camp inmates had died, he might have been tried for war crimes, even though his growing anti-war views had, in 1944, led to a short incarceration by the Nazis. As it was, he was allowed to work for the Americans and, after a short period as a POW, permitted increasing freedoms until he became a US citizen in 1955; his career in space rocketry took him to the very top of the US space programme. Although always regarded with some suspicion because of his Nazi past, von Braun is credited with the Americans' successes, once resources were released in an attempt to overtake the Soviet successes of the late 1950s. A former director of the Apollo programme, Sam Phillips, went as far as stating that without von Braun America would not have been first on the moon.

Although minus von Braun and most of his top team, the Soviets did have several former Nazi programme engineers in their territory, including von Braun's deputy, Helmut Grottrup, in charge of electric and guidance systems. They, and the Soviet's rocket science team, were able to unearth and amass a huge amount of seemingly unconnected information and components left behind and hidden by the Nazis before the war ended. These were ingeniously put together to form

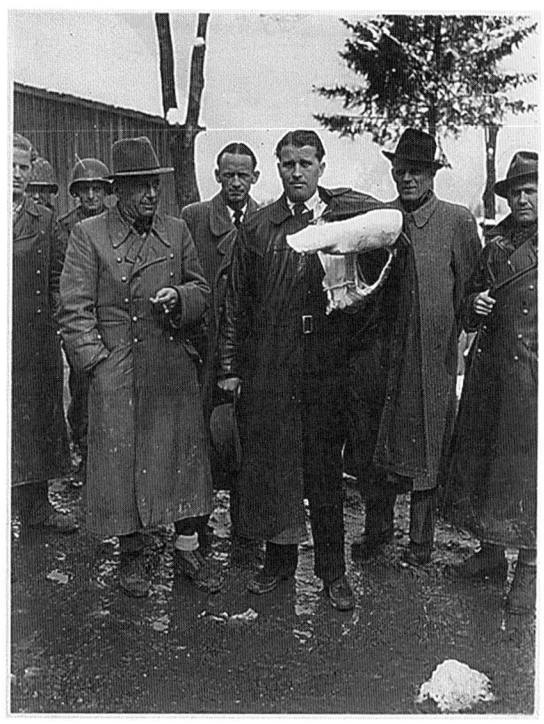

Wernher von Braun after capture by the Americans, 3 May 1945. (*Louis Weintraub via Wikimedia Commons*)

the basis of the initial Soviet missile programme which, in the light of the developing Cold War, had suddenly assumed importance to Stalin and his leadership. It was notable that the Western Allies, in the presence of Soviet representatives, launched an A-4 missile from Cuxhaven, Germany in October 1945. Initially, the Soviets used the facility at Mittelwerk for development (named the RABE Institute) and German sites for initial launches of their own A-4 rockets assembled from parts found and created from scratch. In a typically Soviet version of Operation Paperclip, Operation *Osoaviakhim* on 22 October 1946 saw the overnight swoop and detention (some say at gunpoint) of some 2,200 German specialist engineers and scientists who were moved with their families to the USSR. These included various technical experts not associated with the German rocket programme. In May 1947, 170 of them were taken to Gorodomlya in Central Russia to work for NII-88 (Scientific Research Institute No. 88) established in May 1946 out of RNII to oversee the development of rocket-propelled missiles that could deliver warheads. Korolev was a prominent member of this bureau. Unknown to the Germans involved, they were working in conjunction

The Phoenix that Rose from the Ashes of War 33

Allied launch of V-2/A-4 rocket, Germany, October 1945. (*Wikimedia Commons*)

with another group of German and Soviet scientists in OKB-456 under Glushko, who were developing the engines.

By all accounts and despite the circumstances of their forced relocation, the Germans were well looked after and treated on almost equal terms with their Soviet colleagues. Their tenure, however, was limited, unlike

Soviet A-4 rocket, 1948. (*Wikimedia Commons*)

their counterparts in the US, some of whom (like von Braun) became permanent fixtures in the American programme. By the mid-1950s they had all left the Soviet Union for the West. In later years the Soviets were to declare with pride that while the US had based its space programme on the work of Nazi science and personnel, their own had come from a historic Russian and Soviet interest in space travel – at most a partial truth. The Soviets, however, had certainly limited their use of German personnel to that of maximizing what could be learned from the Nazi programme and once that had been fully scoped, their use had ended.

Soviet Advances in Missile Technology

The combined result of all the Soviet effort with their German experts was the development of the R-2. This was an expanded version of the German A-4 with a new engine designed by Glushko's team, with twice the range – 400 miles. Much of the work done was undertaken at NII-88's highly secret location in Kaliningrad, a Moscow suburb established in 1946 – named in a deliberate act of confusion with the similarly renamed Konigsberg 1,000 miles away on the Baltic coast. These upgraded German missiles, still looking very much like the wartime V-2, were produced in large numbers, and were later said by German scientists who had worked on them, to have a target accuracy of within 1,000 yards by 1950. Stalin was especially interested in the project to develop the next stage – an R-3 rocket with a capability of inter-continental warhead delivery, a project based on unearthed German designs for a two-stage projectile capable of attacking New York from European bases. This project was eventually abandoned due to insurmountable engine problems but led to the first generation of two-stage Soviet inter-continental ballistic missiles (ICBMs) with the successful launch of the R-5 missile a few months after Stalin's death in May 1953, a project that was entirely Soviet in creation. This rapid scale of Soviet developments was largely ignored in the West, with consequences as we shall see in Chapter 4. The R-5 was subsequently developed to become the world's first nuclear weapon carrying a ballistic missile. However, Korolev knew that the long-held interest in space travel he shared with Glushko, Tikhonravov, Chertok and others at NII-88 (renamed OKB-1 – the Chief Design Bureau – in 1956), could become a reality with their rocket developments. All that was needed, in simple terms, was to replace the nuclear warhead with a projectile containing a mechanism that could orbit the Earth and send back information, and the first steps into space could be made.

* * *

No Soviet developments in missile technology were known to the Americans. President Eisenhower had already turned down a request by von Braun and his team to work on space projects. They wanted to use the Jupiter-C rocket they had designed as a larger weapon-bearing ICBM to complement the Thor and Atlas launch systems that were under development by private US corporations. The whole project was put on hold – partly because there was still suspicion of Nazi associations with the rocket team and a wish that breakthroughs of this nature should be all-American. The US administration also shied away from the step Khrushchev had taken – authorizing peaceful scientific development that was closely associated with military technology. The Americans were in for a big surprise.

Chapter 3

Defending the USSR: The Onset of the Cold War and the Arms Race

We have no classes, groups or individuals yearning to seize foreign lands, external markets, or spheres of investment. We have no people who profit by government war orders. In our country no group will ever fan a military psychosis, scaring the parliament into increasing military appropriations and taxes on the population. We have all the resources we need.
N. Khrushchev, at the World Conference for
General Disarmament and Peace, 10 July 1962,
Moscow (Khrushchev 1963, p12)

Although there were ideas, plans and threats, the space race never involved direct military weapon deployment; even Reagan's Star Wars of the 1980s did not proceed far beyond the dreams of its would-be creators. However, for many observers, the race to land a man on the moon reflected military competition at its highest level, short of actual war, and thus an undeclared state of war existed at that time between the superpowers. National prestige and pride were tied up with space achievements, and superiority was interpreted as reflecting military strength – the truth being usually well hidden behind the propagandist showcasing of the 'firsts' – Sputnik, orbit of the Earth, lunar probe, man in space, spacewalk, a man on the moon, and so on. This chapter will provide a brief overview of the Cold War between the USSR and the USA – generally considered to have started in earnest in 1947, and lasting until the demise of the USSR in 1991.

The End of the Second World War

The generally accepted narrative in the West is that only the might and resolve of the US and its Allies prevented the Soviet Union rolling into

The Onset of the Cold War and the Arms Race 37

Western Europe in pursuance of its ultimate goal: spreading revolution and imposing communism after the defeat of Nazi Germany in 1945. The absurdity of this was overlooked by its main advocate, Churchill. His view of the threat from the USSR, expressed in his detailed histories of the period, and in speeches made at the time, has pervaded even the best academic accounts. Only historians with a detectable left-wing bias have challenged the assumptions behind this and presented a post-war history view based more on facts than Cold War rhetoric. None of this is to protect the Soviet Union from historical judgement about Stalin's – and some of his successors' – ruthless methods in dealing with internal opposition, including that expressed in what became the satellite countries on its borders: Hungary, East Germany, Czechoslovakia, Poland, Bulgaria and, less directly, Yugoslavia and Albania. As the war ended, the USSR was tired, scarred, and concerned only about averting the threat of further invasion. Stalin had ceased talk of spreading revolution by the late 1920s and dealt mercilessly with opponents like Trotsky who advocated this without reserve. In 1943 he wound up the Communist International (Comintern) – the last remaining vestige of the idea that international communism might be promoted from the USSR. Stalin's foreign policy was dictated by the need to prevent the growth of fascism as an immediate threat to the USSR's survival, and a wish to consolidate border losses that followed the First World War and the Civil War in Russia. The latter's end in 1921 saw Soviet territory on its western borders reduced by the creation of the Baltic states of Lithuania, Latvia and Estonia, and the expansion of newly independent (and right-wing nationalist) Poland, as it repelled the Bolsheviks in 1920. The fact that these new nations were an expression of the democratic will of their own people was accepted pragmatically by Lenin, but not by his successor Stalin.

The main concern of the Red Army in 1945 was to get home and rebuild their shattered country. Their presence across Eastern Europe had been tactical. The pursuit and defeat of the various German armies and securing the USSR's borders was their paramount strategy, not some sort of chess move to further revolution as has been interpreted by many scholars and historians, exemplified by Hill (2017) whose recent account of the Red Army is regarded as definitive. Slightly more generously, Beevor (2002) ascribes Soviet tactics at the end of the war as seeking to secure as many German POW captives as possible to form a slave army for the rebuilding of the USSR. He, after Churchill, still considers that the Red Army was set on the liberation of Denmark far to the west but were beaten to it by the derring-do of the British Army who, with some German collusion, raced to cut them off east of the agreed limit of Soviet advance on the River Elbe. This claim also seems to lack any grounding – Stalin stuck to the limits of Red Army advance previously agreed with his allies. Towards the end, the Nazis had tried hard to foster division between the Allies to the extent of hinting at peace with the Western nations to jointly defeat Bolshevism. Although this had already been rejected early in the war in favour of a policy of unconditional surrender, the underlying message entered the rhetoric of Western leaders. When Churchill made his famous speech in Fulton, Missouri, in 1946, ruminating about the new 'Iron Curtain' dividing freedom and democracy from the perceived threat from the USSR, he was repeating a phrase spoken by Nazi propaganda chief Goebbels in 1944. In 1946, the future of the nations occupied by the Allies at the end of the war remained undecided. Bulgaria, Rumania and Hungary had all been German allies and had taken part in the invasion of the Soviet Union – all had changed sides towards the end – and it was to be a few years before each became countries with regimes similar and sympathetic, if not subservient to the USSR. Germany was still full of Nazis, and efforts and initial enthusiasm for rooting them out were soon sidelined by the West who realized that lower-ranking (and some senior) former Nazis were needed to run the country if Germany was ever to recover to the point of self-sufficiency. Stalin showed his conciliatory and peaceful intent by moving his armies out of territory occupied in Austria

Victory Parade, Red Square, Moscow, July 1945. (*Wikimedia Commons*)

and Finland – the latter based on an agreement that guaranteed it would never again be used to attack the Soviet Union. It does seem that the evidence for a Soviet plot to take over Western Europe at that point is rather far-fetched.

This wartime history is important because the roots of the Cold War are generally believed to involve Soviet behaviours during the war – the truth perhaps is that these were perceived for ideological reasons rather than actual ones. At the end of the war, with Europe in ruins, and its weary citizens hoping, and expecting, a better world than the one many had experienced before 1939, the threat of communism was

'We Demand Peace', a 1950 poster by Viktor Koretsky. (*Hi-Story/Alamy Stock*)

'The Missile Shield Will Surely Defend Our Peaceful Work', 1964 Koretsky poster. (*Hi-Story/Alamy Stock*)

believed to be great, whatever ideas Stalin may have had. In fact, he stood back when the British quelled a popular left wing-led rebellion involving communists in Greece in late 1944, but manoeuvred to ensure that the nationalists in Poland were suppressed in favour of his supporters. The difference was that Poland was on the USSR's border (and within the sphere of influence agreed with the Allies) and there had been constant friction through the 1920s and 1930s with its right-wing government. In Britain the Communist Party was outdone from the left – the Labour Party swept to power in June 1945 on a programme of nationalization and the promise of a welfare state. The idea that the Soviet Union threatened democracy suited conservative elements in Britain, the US, and elsewhere. Turning the USSR into a new threat suited their efforts to stifle socialist views at home. Many in the ruling establishment had been uneasy about the wartime alliance; Churchill, its staunchest advocate, always viewed it pragmatically and was quick to revert to the anti-communist position he'd held since the 1917 Russian Revolution once the war's end was in sight. One popular US book from 1946 reprinted the entire theses, statutes, rules and programme of the Communist International dating back to the early 1920s, with an opening chapter arguing that these officially obsolete documents still represented the Soviet Union's real agenda (Chamberlain 1946). This trend soon moved beyond all reason – US McCarthyism of the late 1940s and 1950s excluded those under suspicion from civic life and employment. It was a paranoid reaction to fabricated threats that deliberately sought to characterize all left-wing sympathizers as crypto-communists and sworn enemies of the nation. The reality was pointed out by Frank Roberts, Britain's ambassador to Moscow, in 1946:

> Although Soviet Russia intends to spread her influence by all possible means, world revolution is no longer part of her programme, and there is nothing in the internal conditions within the Union which might encourage a return to the old revolutionary traditions.
> (Quoted in Hobsbawm 1994, p225)

The only bit missing in this statement is that Stalin would happily leave his supporters high and dry, as he did in Greece between 1944 and 1948 if it suited his broader foreign policy strategy.

Early Moves in the Cold War

The US government under Truman (from April 1945) and his successors were less honourable than Stalin when it came to peace. Truman tried unsuccessfully to use America's ownership of the atomic bomb to force concessions from the USSR that reneged on earlier wartime agreements – these included the suggestion that Red Army troops withdraw to the Soviet Union's borders. Stalin would have liked American help with rebuilding his country but once it became clear that conditions for this amounted to the acceptance of capitalism over the socialism, he stood firm. This was a surprise to the Americans at the July Potsdam Conference once they had told Stalin about their new weapon. Their response was to ramp up anti-communist rhetoric. Stalin also stuck to his 1943 promise to join the war against Japan within three months of the conflict ending in Europe. This disappointed the Americans who would have preferred victory on their terms with the use of the atomic bomb – keeping Soviet influence out of China and the Far East. This confirmed Stalin's belief that he could only secure the USSR's borders through the establishment of friendly regimes in the neighbouring countries – meaning a longer journey for American bomber fleets who might attack Soviet cities. Tensions inevitably increased over the next few years and US industrialists (the 'Military Industrial Complex' as they were described), who had made massive profits from war production, were delighted that a new enemy provided a further opportunity to continue high levels of arms spending, for the foreseeable future, as the US sought to replace the old colonial powers, like Britain, with economic and ideological dominance over the post-war world.

Until the end of the war, the Soviet Union's build-up of its armed forces had pragmatically focused on the defeat of the Germans rather than an arms race to head off potential aggression from the enemies of the future. It took several months after the victory in Europe before sufficient forces could be organized to move thousands of miles to the Far East to join the war against Japan. The post-war world brought entirely new challenges – the threat came from a power who had bases and allies across the globe, and close to most of the USSR's vast borders. This required the development of military strength that might equal that of the US and other potential foes like the United Kingdom. Not only would the USSR have to develop a nuclear capability, but it would also need to build its air and naval forces and maintain its vast army across the buffer states under its influence. For an economy shattered by war, and within which corporate profit had no place as a driver of growth and prosperity, this was to prove, in the long term, an impossible task.

In 1947 US President Harry Truman made a foreign policy announcement to Congress: the Truman Doctrine. This was aimed at providing immediate American support for forces combating communist-backed popular uprisings in Greece and Turkey. Surrounded by dubiously based rhetoric concerning democracy versus totalitarianism, it was soon extended to embrace US foreign policy with a commitment to combat Soviet influence anywhere in the world. One of America's first moves to isolate their new enemy was to create a military alliance, NATO, in 1949, whose avowed aim was to counter the threat of Soviet attack on the West. This effectively brought the military forces of allies like Britain and France (who later withdrew for a period) under US control and leadership (as the largest and most powerful contributor). The other component of American policy was the Marshall Plan, officially known as the European Recovery Program, which sought to head off the spread of communist influence, through the investment of US dollars, in participating countries to develop productive capacity and trade. The Soviets, who were invited to participate, correctly saw it for what it was – a plan to reinforce American capitalist values at the expense of socialist ones. They rejected it for themselves and the countries of Eastern Europe under their influence.

The Arms Race, MAD and Peaceful Coexistence

It seems that it was more by chance (or accident) than probability, that the world survived the Cold War without destroying itself through thermo-nuclear war. The Americans went onto full nuclear alert on nineteen occasions – the Soviets on only one (against the UK and France at the time of the Suez Crisis in 1956). This fact gives away the truth about the Arms Race – that the Americans were always ahead and the Soviets always had to back down when it came to confrontation. This they did on several notable occasions, including Cuba in 1962. However, the USSR substantially developed its armed forces after the Second World War.

By 1949, the Soviets had exploded their first nuclear bomb, built from intelligence gained through their spy rings located within the heart of the wartime Anglo-American Manhattan Project. Until Stalin learned of the nuclear test success in 1945, the Soviets' own programme had been very small but was expanded quickly to match their opponents. From then on, the Soviets played a catch-up game with the Americans until the 1980s – always lagging in terms of numbers, hydrogen bomb development, explosive power, ICBM capability, multiple warheads and eventually submarine-carrying capacity. Despite this unequal competition, both sides soon developed sufficient stocks and means of delivery to destroy humanity several times over. The production of nuclear weapons leapt forward in the early 1950s with the thermo-nuclear breakthrough by American scientists that allowed vastly more powerful hydrogen bombs to be packaged into a small bundle that could fit into a missile nose cone. MAD – mutually assured destruction – assured that no side would want to start a conflict that both would ultimately and quickly lose, but there was a very real threat of accidentally triggering a war. In terms of numbers, the Soviets only ever possessed a fraction of the warheads available to the US. At the time of the Cuban Missile Crisis the Americans probably had about eighty, and the USSR four that could be reliably targeted – if such odds even matter in a suicide pact. By the 1980s, this ratio had narrowed but had reached fantastic proportions – the US had some 9,000 warheads (to which could be added those of its British and French allies) to the USSR's 4,500 odd. The idea that the threat came exclusively from the USSR, as governments in the West continued to argue, was never accurate.

Soviet arms expenditure was not just about nuclear capability. It reduced its army from some

An American 1962 atomic bomb test from a US submarine periscope. (*US Navy via Wikimedia Commons*)

twelve million to (a still massive) three million within a few years of the war's end but the Cold War ensured that it maintained a military capacity to match that of the US. The navy, whose role in the Second World War had been peripheral, grew to become the world's second-largest fleet. It included surface ships and submarines capable of global operations. Even so, it always remained small in comparison to the US Navy. NATO only started taking the Soviet Navy seriously when they developed long-range nuclear missile-carrying submarines in the early 1960s that could match the Americans' Polaris submarine fleet. The Soviet Air Forces acquired jet fighters soon after the war and, ironically, their MiG-15 fighters could outperform anything the US possessed in the skies over Korea – powered by improved Rolls-Royce engines the British had arrogantly believed could not be copied when they sold them to the Soviets in 1947. ICBM expansion was pragmatically developed to provide the nuclear shield as a cheaper option to aircraft bomber fleets that were expensive and vulnerable to interception. However, long-range bombers were developed to replace the obsolete fleet of TU-4 aircraft that were exact copies of the wartime US B-29 bomber. The TU-95 'Bear' was the most-produced and longest-serving design – often photographed from NATO aircraft as it made trips over the Atlantic. Air defence strategies also involved the rapid

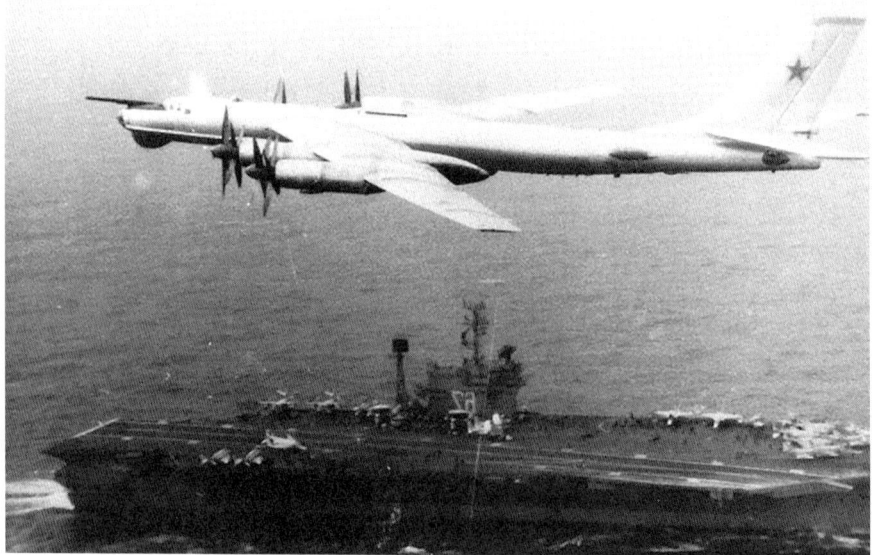

Soviet TU-95 'Bear' overflying USS *John F. Kennedy*, Atlantic Ocean, April 1969. (*US Navy via Wikimedia Commons*)

development of radar technology post war. Stalin had been impressed by the manner in which Allied bomber fleets had flattened German and Japanese cities and the nuclear age lent urgency to catch up. Fairly effective systems were eventually deployed, based initially on German technology and expertise, but the area to defend was so large that their coverage was always a matter of speculation to Western observers.

By the late 1950s the US, and their nuclear-armed allies, had developed very fast low-flying medium bombers that could avoid radar detection and fly quickly from their European bases to launch nuclear missiles on targets in the USSR. The Soviets struggled to keep up with these advances. Indeed, they argued consistently for 'peaceful coexistence' and for disarmament talks to commence.; these were rejected by the Americans on the basis that they were wholly dishonest. However, one British expert on Soviet aviation from the time (Lee 1961) believed that the USSR's proposals were entirely genuine – based on pragmatic realities and the fact that more and more countries were acquiring such doomsday weaponry. Even the more confident rhetoric of Khrushchev in his speeches in the early 1960s reflect his desire to end the arms race. This was also apparent in Soviet actions. From Stalin's time, all new housing in Moscow and other major cities in the USSR had to have underground bomb shelters. This was abandoned on Khrushchev's orders in 1957, as civil defence measures in the US were being stepped up as a result of Sputnik 2's success. So paranoid were some US leaders over the threat of Soviet attack through their presumed technical superiority, that an accelerated programme of nuclear bomb shelters was demanded, and refused by a cost-conscious President Eisenhower.

Ground forces remained the subject of Soviet investment, their hardware being studied closely by Western observers at every Moscow military parade. Where the USSR failed was in the development of information technology – the digital revolution from the 1980s required new ways of thinking that the immobile Soviet system, with its ageing leaders, seemed incapable of mastering, despite a special ministry created for the purpose. This impacted on the military too, crucially, on the eventual outcome of the space race. Meanwhile, the closing of borders and the creation of the Iron Curtain prevented free movement – both of people and ideas. In Berlin, this divided the city quite dramatically when the wall was built in 1961, although some observers have said that East

A US military photo of a Soviet nuclear warhead carrying ICBM, Red Square, Moscow, October 1965. (*CIA via Wikimedia Commons*)

Berlin Wall, US–Soviet military stand-off, October 1961. (*US Army via Wikimedia Commons*)

German and Soviet troops were under orders to avoid confrontation and might well have backed down if the Western powers had simply knocked the wall down as it was being built.

On several occasions, global conflict was only narrowly avoided. In 1962 Cuba invited the Soviets to site ICBMs on its territory – a move that replicated the location of US nuclear missile bases in Italy and Turkey, close to the USSR. Cuba had only recently replaced its corrupt pro-US dictatorship with a popular socialist regime that had won power through a successful revolution and armed struggle. As it was only ninety miles from the US coast, this was a source of immense irritation to the Americans and they had backed an attempt to topple the new government with an easily repelled invasion at the Bay of Pigs in 1961. The siting of the Soviet missiles, soon detected and made known to the world, was considered by the US to be a major act of provocation. Khrushchev and US President Kennedy squared up to one another diplomatically and militarily. On one occasion a Soviet submarine, the *B-59*, near the Cuban coast and the US naval embargo line, was depth-charged with practice bombs by the frigate USS *Beale* to bring it to the surface and find out its business. The Soviet commander was under orders to retaliate and attack with his single nuclear torpedo but was so deep in the ocean and out of radio communication that he had no idea what was happening and whether a war had broken out. He wanted to launch the torpedo but required the consent of two other senior officers aboard, and one, Captain Vasily Arkhipov, refused to give this, believing that it would trigger a conflict and that further orders from Moscow were required before any action was taken. The submarine surfaced, became aware of the real situation and then left the area without incident. On the same day an American U-2 spy plane, whose navigational systems were confused by the Aurora Borealis, accidentally strayed over the USSR from Alaska. The Soviets, fearing an attack, deployed jet fighters to intercept and destroy the plane, which in turn triggered the deployment of nuclear-armed American fighters. Thankfully, the U-2, which had now run out of fuel, was glided back into US air space before the two sides met, and conflict was very narrowly averted. The next day the two shaken leaders agreed to talks and a secret compromise agreement was eventually reached which saw the Soviets abandon its Cuban missile bases in return for the withdrawal of nuclear missiles from US sites in Turkey and agreement not to back any further invasion attempt of Cuba.

Prospects for Space Wars

As the space race intensified in the late 1950s, both sides looked at proposals for the siting of weapons and anti-missile defence systems in space. On the Soviet side, Korolev had little interest in such matters, being concerned with more basic space discovery and experimentation. However, another rival scientist, Vladimir Chelomey (1914–1984) was appointed chief designer of aviation equipment following his work on developing ICBM technology. His early space proposals were ambitious and rejected by Korolev, but their respective bureaus were combined after Khrushchev's removal in 1964 and some of Chelmoney's designs were incorporated in the lunar space programme. In 1960 he and Korolev had a meeting with Khrushchev where Chelmoney put forward proposals for a military space station and an anti-missile system not unlike the US Star Wars ideas of later years. Khrushchev had little time for what was regarded as fantasy and they did not proceed – although Chelmoney's new ICBM proposals were received more favourably. Nothing more was heard about this and the only reference to such discussion is in the memoirs of Khrushchev's son, Sergei (quoted in Siddiqi 2000, p234–35).

Meanwhile, at around the same time, the Americans were more openly looking at the prospect of war in space. Some of the popular publications from the late 1950s are quite chilling in the open discussion of what leading figures in the US military establishment believed might happen. The surprising Sputnik successes (discussed in Chapter 4) made these people feel very sore and quite threatened:

> The United States must win and maintain the ability to control space in order to assure the progress and pre-eminence of the free nations. If liberty and freedom are to remain in the world, the United States and its allies must be in a position to control space. We cannot permit the dominance of space by those who have repeatedly stated they intend to crush the free world. (General Thomas D. White, Chief of Staff, United States Air Force, March 1958 (in Straubel et al., p11))

An influential and well-respected US general, James Gavin, retired from the military to write a book (Gavin 1959) condemning the US establishment for downplaying the communist menace, the means needed to beat it and, in particular, its lead in space technology. He predicted that by 1965 wars might be fought by tactical nuclear weapons guided and launched from satellites in space, and bases on other planets. For Gavin, the Third World War started in 1945 and was being fought everywhere except the home territories of the USA and USSR. He suggested that as the conflict was global, mobile US missile systems aimed at the USSR should be based in Africa (he nowhere suggests that this might involve a requirement for the consent of its populations). Thankfully, Gavin's predictions did not come to pass.

One of Gavin's main concerns was that the Soviets were winning the psychological aspect of his Third World War. At the time, they were leading in the space race and their approach to nuclear disarmament seemed more peaceful than proclaimed US intent. Their proposal for a nuclear test ban, which Gavin was very much against, succeeded a few years later in international agreement while US aggression in Vietnam, and interference in Chile and other parts of South America, was leading to criticism at home and abroad. The Soviets, on the other hand, preached peace and mutual coexistence. Their support for liberation movements in Africa and elsewhere won applause from people across the world who had nothing to thank the former colonial powers for – the same powers who were the USA's principal allies in Europe. Gavin had good reason to believe what he did and was proved correct as events unfolded. However, the Americans, as we shall see, eventually won this war. Gavin died in 1990, no doubt a happy man at seeing the disintegration of the USSR.

With neither side capable of the expensive, and as yet out-of-reach technology, required for space wars, it was in the interests of both that agreement should be made about eradicating such threats. The result was the Treaty on Principles Governing the Activities of States in the Exploration and Use of Outer Space, including the Moon and Other Celestial Bodies (known as the Outer Space Treaty) signed in 1967 – initially by the USSR, USA and United Kingdom. Since then over a hundred other nations have signed too. This is a simple declaration that opens up outer space and other planets for free exploration and use, outlawing declarations of sovereignty over the moon or any other celestial body, and the siting and use of weapons of mass destruction. What was not prohibited was the use of space for military surveillance, communication and weapons control systems. By the time of Gary Powers's conviction in

April 1961 (see Chapter 4) the U-2 spy plane was already being replaced by US Corona Program satellites that would eventually be capable of providing detailed photographs of Soviet military facilities from space. The Soviets, as with much Cold War technology, struggled to keep up with these developments but did have their own programmes.

Cold War Propaganda

The military build-up was accompanied throughout the Cold War period with a war of words. The US authorities promoted various mediums to convince the population that communism was evil and against human nature. Some of these, such as the film documentaries that were intended as serious viewing, look quite naïve and simplistic today – mirroring the Captain America stereotyping for young readers. *The Challenge of Ideas* (1961), designed for US servicemen of ordinary rank who were destined for service on the frontline against the Soviets in Europe, was narrated by a plain-speaking, cigarette-smoking Ed Murrow, a well-known broadcaster from the Second World War era. It included an interlude from the movie star John Wayne who reminded viewers of the primacy of the 'American way of life'. A more sophisticated film from 1964, *What Everyone Should Know About Communism*, was presented by an 'expert' who at one point adopts a pseudo-Russian accent to describe the world from the Soviet view, switching back to his native voice to condemn all he had just said as persuasive, but highly dangerous, half-truths.

The 22nd Congress of the CPSU in 1961 was full of invective against the evils of capitalism and the superiority of the Soviet Union and its friends. As we shall see in later chapters, over and above the proud trumpeting of scientific achievements in space, no opportunity was missed to compare the two competing systems. Soviet films, posters and every kind of education for younger people extolled the peaceful intent of the USSR and the imperialist and profiteering motives of their adversaries. Such crude Soviet propaganda has helped form the popular view of the era of real existing socialism – especially in the narratives of many of the former socialist countries outside Russia. What is not so well known is the extent to which similar, but less obvious, efforts were expended in the more liberal West to negatively characterize its Cold War enemies. A 1985 study of British school textbooks about recent history concluded: 'The result is that children are being given a view of 20th-century history which a wide body of historians would consider *distorted* or *extreme*' (Sykes et al. 1985, p39). This concerned accounts of the end of the Second World War and the start and continuation of the Cold War, which it felt unreasonably blamed the Soviets for deteriorating relationships and ongoing conflict.

Statements by both sides were open to interpretation that suited the purposes of the listener, none more so than Khrushchev's 1956 statement to Western leaders: 'Whether you like it or not, history is on our side. We shall bury you.' This was widely quoted afterwards by Western leaders and generals to justify their war plans. Few heard a subsequent Soviet explanation that this was a reference to Marx's 19th century dictum, that capitalism was its own gravedigger and that replacement by communism was an inevitability.

Reagan, Gorbachev and the Final Period of the Cold War

Between the 1960s and early 1980s, a situation of understanding over MAD existed between the USA and USSR, and the Soviet notion of 'peaceful coexistence' prevailed at an international level. By this time aircraft-borne nuclear weapons had been replaced by ICBMs of various types, including significantly, almost undetectable submarine launch systems. Détente was confirmed by the Strategic Arms Limitation Talks (SALT) of 1972 which limited development in particular areas and helped maintained this dangerous balance. To avert war a telephone hotline linking the White House and the Kremlin had been put in place in 1963. Both sides, as they had in the Korean War in the early 1950s, pursued military means to support allies in the Third World who were involved on one side or another in struggles to free themselves from colonialism or the reactionary regimes that had inherited government. The US was certainly the most belligerent in this respect, the awful war in Vietnam, involving its neighbours Cambodia and Laos, being the most notable example. This involved three times as many bombs being dropped from US aircraft than in all theatres of the Second World War. For their part, the Soviets supplied the North Vietnamese with increasing numbers of very effective anti-aircraft defence systems and fighter aircraft, without openly joining the battle. The Soviets eventually became bogged down in

Vietcong fighters during the Vietnam War. (*Bộ Quốc phòng via Wikimedia Commons*)

their own Vietnam – Afghanistan – in the 1980s (see Chapter 9), which, like the Americans (who similarly armed their opponents), ended in Soviet withdrawal and residual conflict that continues to this day.

By the late 1970s, with renewed tension created by a worldwide recession, the stand-off concerning MAD was reinforced by the development of systems by both sides that could provide effective (or so it was hoped) early warning of attack. This in itself was problematic, it was open to error and on a notable three occasions, war almost resulted.

The first such incident was in November 1979 when a technician in the North America Aerospace Defence Command (NORAD) accidentally set off a training programme simulating a Soviet missile attack. This was taken seriously – intercepting aircraft were mobilized and the president's Doomsday Plane, in which he would sit out a nuclear attack, was ordered to take off. However, further checks revealed the mistake and stand-down followed. The two sides exchanged terse correspondence after this but similar, if less dramatic, alerts took place at NORAD due to microchip failures in 1980. An erroneous warning at the USSR's equivalent site in Serpukhov mistook rays of sunlight over the US ICBM launch site in Montana as an attack by five missiles. The commanding officer, Colonel

Petrov, was sceptical and held off alerting the high command of an attack for several tense minutes until confirmation was received that the alarm was false. Again, a nuclear holocaust was narrowly averted.

In 1980 Ronald Reagan was elected to the US Presidency. He was unhappy at détente. Reagan believed the Soviets had to be beaten back by US military might and set about on a new armaments programme that would outpace Soviet ability to compete and, by this logic, force an end to the game. Part of his strategy entailed a fresh ideological campaign in which, borrowing language from the blockbuster movie *Star Wars*, he described the USSR as an 'evil empire'. Pershing cruise nuclear missiles were established in Europe, including the UK, and the Cold War increased in intensity, mobilizing a vast peace movement across the Western world. The new breed of missiles (including the Soviet SS-20 and Backfire-B bomber) were mobile and could be precisely targeted. These were aimed not at cities, but military targets – missile silos and related system locations, that made the danger of the first strike – to knock out the other side's offensive capability, a very dangerous concern. In November 1983, Exercise Able-Archer, a large-scale NATO exercise along the borders of the Eastern Bloc was mistakenly believed by the Soviets to be the start of an invasion of East Germany. Their forces across Europe were placed on high alert but stood down when the exercise ended and the false alarm was realized. At the time the US and their Western allies were unaware of how close they had come to war. It should be remembered that in 1983 the USSR, who were defending a border against the Chinese as well as being involved militarily in Afghanistan, only had some twenty divisions in Europe. They had deployed 400 to defeat the Nazis in 1945, so a nuclear defence option was a very real danger. Reagan announced the Strategic Defence Initiative (SDI) programme in 1983 which involved the development of defence systems that could destroy ICBMs in the air by ground, air and space satellite-based means including lasers and missiles – the so-called Star Wars programme. This, as it turned out, was mostly bluff and bluster, but millions of dollars were poured into research and development until it was abandoned (for a little while at least) in the wake of the end of the Cold War.

Reagan's confrontational and risk-laden strategy was essentially successful. The new Soviet leader, Mikhail Gorbachev, was attempting to save the Soviet Union from total collapse. Internal tensions over economic

US President Ronald Reagan addresses the nation on the Soviet threat, 1985. (*US National Archives via Wikimedia Commons*)

problems were mounting, and the levels of military expenditure required to try and keep up with the US were simply not available. Initial talks in 1986 and early 1987 were not productive as Reagan refused to stand down his Star Wars programme. However, in late 1987 and 1988, after Gorbachev announced a reduction in Soviet nuclear weapon deployment, further talks resulted in agreement on the mutual reduction of arsenals and troop deployment in Europe. By 1989, the USSR was in its death throes with the dismantling of the Berlin Wall and effective removal of

Soviet ballistic missile-carrying submarine, 1968. (*CIA via Wikimedia Commons*)

Soviet missiles return to the USSR from Czechoslovakia as a result of disarmament talks, 1987. (*Vladimir Rodionov RIAN Novosti via Wikimedia Commons*)

the Iron Curtain that had been in place since the early 1960s. Despite this, the US military establishment sought to characterize the Soviet Union as a danger right up until the end. The US Defense Department's annual review for 1990, *Soviet Military Power*, continued to argue for nuclear weapons and continued high spending by the West. The Cold War ended when the USSR's CPSU renounced power and stood down as it was declared illegal in 1991 – matters discussed in Chapter 9.

* * *

The cost to the Soviet economy of the Cold War was enormous. From the 1960s, military spending rose, according to US official estimates, by a planned 4–5 per cent per year at a time when the economy, measured in terms of gross domestic product (GDP), was in decline. From a peak of 20 per cent of GDP in the early 1950s and maintained at a high level up until the 1980s, it had fallen to about 10 per cent by the end of the Soviet era. In the same period US arms spending, which far outstripped that of the Soviets (except for a short period in the early 1980s), represented 12 per cent of GNP reducing to 7 per cent by 1990. It was only by allocating a greater proportion of GDP to arms spending that the USSR was able to keep up with the West throughout the Cold War period. In almost all other areas it was behind – from the personal incomes of the majority of citizens to technical superiority in most fields of science, manufacturing and agriculture. The Soviet Union, for example, was never able to adequately feed its population (see Chapters 8 and 9). By the late 1960s, the Americans and Soviets were estimated to be spending $50 million a day on their nuclear armaments each – a figure ruinous to the USSR. It was clear that as a richer country the US was always going to win – a reality eventually accepted by Gorbachev.

Despite the awful human cost and destruction via the endless proxy conflicts, the fact is that neither side ever made a move against the core territory of the other. The aggressive war of words was always mainly conducted by the Americans – the Soviets generally preached a message of

Berlin Wall, 1986. (*Selbst Fotografiert via Wikimedia Commons*)

The Onset of the Cold War and the Arms Race 57

Gorbachev and Reagan meet, 1987. (*US Government via Wikimedia Commons*)

Disabled Soviet Badger bombers, late 1980s. (Soviet Military Power 1990 *via Wikimedia Commons*)

peaceful coexistence and a wish for mutual disarmament. After the Cuban Missile Crisis, a leading Soviet official stated that such a climbdown, forced by the Soviets' relative weakness, would never happen again (he assumed that the Soviets would catch up with the arms race) but it did, although such aggressive pronouncements were rare. Stand-offs took place and enormous sums of money were spent, but a direct conflict was always avoided. The places in the world outside the Eastern Bloc where the USSR had influence were, by the 1980s, few and of little strategic importance: Angola, Mozambique, Ethiopia, Cuba, Afghanistan and South Yemen. Some of these, such as heavily subsidized Cuba ($6 million a day) were costly to the Soviet people and determined similar spending to promote pro-Soviet regimes – Afghanistan proved to be especially disastrous (see Chapter 9).

Finally, having matched itself against the US for over forty years, the character of its adversary was indelibly imprinted upon the people of the declining Soviet Union – including its leadership. What the US achieved became the yardstick for USSR success, and its consumer items the envy of Soviet citizens. When the edifice crumbled in 1991, it was no accident that many in the regime's upper echelons sought to grab what assets they could in a frantic American-style goldrush, making vast fortunes for themselves in an ideological leap to the free market (see Chapter 9). That perhaps is the true legacy of the Cold War for the people of the former Soviet Union.

Chapter 4

First Steps to the Stars: Sputniks and Canine Scouts, 1956–1960

We in the United States can be first. If we do not expend the thought, the effort and the money required, then another and more progressive nation will. It will dominate space, and it will dominate the world. There is a nation with this ambition. We must not let it prevail!

James H. Doolittle, former US Air Force general,
March 1958 (in Straubel et al. 1959)

Technological advancement never happens in a vacuum, just as science is never neutral – developments occur as the result of the needs and political requirements of the state, and increasingly the private agendas of corporate donors. In the 1950s and 1960s, it was the needs of the richest states that defined human progress; their actions were often driven by the progress of the Cold War. This chapter will describe the early days of the Soviet space programme and why and how it became a priority for a nation who had always struggled to meet the most basic needs of its people.

The Space Programme Becomes Official

On 25 February 1956, Nikita Khrushchev made a speech, in closed and secret session, to the 20th Congress of the CPSU. It was soon leaked and was made known throughout the USSR and the world. It was published in its entirety by the US government in June of that year, shattering the socialist world. Khrushchev, who had been quietly dismantling Stalin's repressive apparatus since assuming leadership after his death in 1953, detailed at length Stalin's crimes against his people and the 'cult of the individual' he had built around himself. This galvanized opposition both at home and in the satellite countries. Some of this, as in the Soviet republic of Georgia, was pro-Stalin in nature; elsewhere, in Poland and notably Hungary, opponents of the regime who wanted more democracy and self-determination took this as an encouragement to protest and make further demands of the regime. Regardless of motive, all were put down with force – in Hungary 2,500 citizens and over 700 Soviet troops were killed in what had become an armed uprising. This led to a crisis in communist parties throughout the world, with much soul-searching and defections taking place: a gift to international Cold War adversaries. At the moment Khrushchev tried to lead the USSR to a more enlightened future; he was responsible for oppression that seemed to demonstrate that nothing fundamental had changed. Within the Soviet leadership he was under pressure from those who were unhappy with his denigration of Stalin. All in all, he desperately needed to project Soviet success and achievement at home and abroad.

The Soviet economy at the time of Khrushchev's ascendancy was not in great shape. The requirement to keep up with American technology and spending on weaponry was taking its toll. By the mid-1950s, the Soviets were spending at least a third more than the US on defence, and these differences were to be accentuated dramatically by the end of the 1970s. Many Soviet people, and not just those in areas badly damaged in the war, lived in primitive and overcrowded conditions, and food was in short supply. Khrushchev had plans to overcome these deficits: the Virgin Lands Campaign to boost agricultural production was one (see Chapter 8). Mass housing construction was another. In defence, as described in Chapter 3, missile development as a means to deliver nuclear warheads was seen as cheaper than trying to build the mass bomber fleets that the Americans were still reliant upon.

Korolev, who was always absorbed in his work, probably cared little about any of this beyond the need to develop ICBMs and the support it resulted in for his space project. The Americans were more open about their plans and Korolev was aware that the US space exploration programme was being quietly promoted by von Braun in the same way that he was courting support in the USSR. The US Jupiter-C rocket eventually penetrated the Earth's upper atmosphere with two-stage launches in August 1956 and then again in May and August 1957. These, however, were military test flights with ICBM advances in mind but demonstrated potential. Korolev knew of these tests and misinterpreted them, believing that the

Americans only required a third stage to launch a satellite into space, and were about to do so. The American view of the Soviet Union at this time was fairly pejorative, as suited the Cold War and anti-communist narrative. It was assumed that in a country which struggled to provide its people with anything beyond the basics, that technological advance was either bought or stolen. Security was tight and what knowledge there was of Soviet advances, was gleaned by studying photographs of what was on display at the May Victory Parade in Moscow. As time went on, U-2 spy plane observations filled in some of the gaps, but in 1956/7 this lay in the not very distant future and nothing, beyond rumour, was known about Soviet space plans. The Americans felt under no pressure.

While Kaliningrad (Moscow) was the location of the USSR's rocket and missile development and assembly processes, a site was needed for launching. The existing steppe launch site at Kapustin Yar, east of Stalingrad (now Volgograd) in southern Russia, had been used since 1946 for launches of the German-derived missiles but was not suitable for the larger rocket developments now underway. It was also close to American intelligence radar installations in Turkey. A commission involving Korolev selected a remote desert location near the settlement of Tyuratam in the Soviet Kazakh Republic, far from prying eyes and, importantly, on a flat plain where radio signals could be received and sent without interruption. It was also close to the Earth's equator so that the effect of the Earth's revolutions was minimized. The site was one of the most expensive infrastructure projects ever embarked upon in the USSR, involving not just construction. The town needed to serve its needs and house its staff, as well as hundreds of miles of railway track from the conveniently located Moscow–Tashkent main line. Opened in 1955 for ICBM testing, it was soon diversifying into Korolev's space programme. The new base was named Baikonur in 1961 and soon became known across the world. This was a necessity, as the siting of Gagarin's launch had to be made public if it was to gain officially recognized international status. It is alleged that a junior officer was tasked with finding a name and simply chose at random a small mining town 200 miles away so that the actual site could not be easily identified. By that time the Americans had already photographed the launching base from a U-2 spy plane. Leonov (Scott & Leonov 2004) later described the site as an endurance test in itself: it was beset with lethal scorpions, snakes and spiders. Accommodation in the

62 Soviets in Space

The Baikonur Cosmodrome as seen from a U-2 spy plane in the late 1950s.
(*CIA via Wikimedia Commons*)

early years was built in accord with Moscow specifications and failed to cope with the sub-zero 40°C cold in the winter and 40°C-plus dusty heat in the summer, all accompanied by incessant high winds.

The R-7 Rocket and Sputniks: Reds in Space and Red Faces on Earth

The R-7 rocket (nicknamed 'Семёрка', pronounced 'Semyorka', which translates as 'old number seven') was first conceived in 1953. As a much larger missile than the previous R-5, it took several years to develop and overcome problems of engine design, guidance and other issues related to its sheer size and range. In its conception, it was ahead of American designs for ICBMs in that the USSR's requirement was for a missile that could send a warhead a very long distance, very quickly. With their bases so much nearer to Soviet soil, the US had less need at this stage. However, soon after the initial development of the R-7 began, the Soviets

First Steps to the Stars: Sputniks and Canine Scouts, 1956–1960

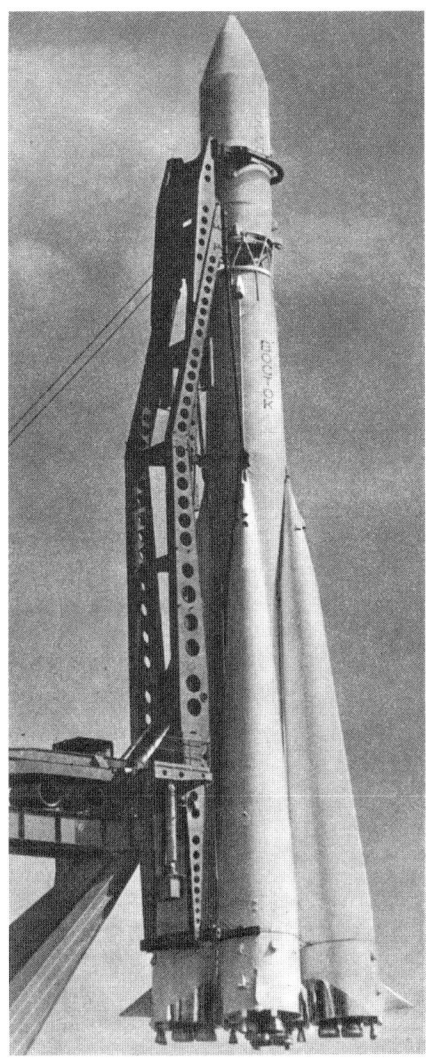

Soviet R-7-powered Vostok rocket. (*Novosti* USSR Probes Space *1968*)

caught up with the thermo-nuclear advances that made warheads substantially lighter, and although it was no longer crucially required as a launcher for weapons, it was deployed as such for several years once past development stage. The R-7 was, however, eminently suitable for space exploration and was developed by Korolev and his team of space enthusiasts with this purpose very much in mind. Its uniqueness was in its use of four booster strap-on engines that would help propel the missile into space and then get jettisoned, propulsion then being taken over entirely by a core motor. The warhead section could easily be replaced with a capsule in which could be located any variety of satellite, including versions that could carry a human into space. The idea of such multi-stage rockets had been examined by Tikhonravov in the late 1940s, with space and satellites very much in mind, but had been ridiculed at the time. He persevered and the later successes of the Sputnik programme brought him belated recognition as a core member of Korolev's team.

In 1954, Korolev and Tikhonravov won permission from their political chiefs to develop rocket and satellite technology with a clear aim of eventually placing a manned version into space. Much of this was based on a detailed and far-seeing report produced by Tikhonravov. Interest was spurred by widely publicized intent on the part of the US to place a satellite in space in 1957 – a year designated International Geophysics Year because of the expected intense solar activity. With the Soviets lagging behind the US in many prestigious areas of science and technology, Korolev persuaded Khrushchev, in a meeting in February 1956, to agree that his team should prioritize a space launch as a peaceful project to beat the Americans and trumpet Soviet engineering and science. However, Khrushchev made it clear that this would take second place to the ICBM weapon system with which it was closely associated – given that the R-7 could serve both requirements. Khrushchev had been amazed at the size, speed and range of the R-7 rocket, especially with its potential ability to attack American cities within a fraction of the time and at less risk of interception than any system then available. With the added prestige of being the first to get into space, it must have seemed a win-win for the USSR. However, the promise of quick success based on little more than a mock-up missile was premature, and immediately placed Korolev and his team under immense pressure, especially because of the mistaken belief that the Americans were much further along in the field of satellite and rocket technology than they were.

The R-7 was completed in double-quick time and was ready for its first test launch on 15 May 1957. This and a subsequent test on 11 June both failed (all hidden from public knowledge). However, a third test, on 21 August, succeeded in transporting a dummy warhead 3,700 miles into the Pacific Ocean. The Soviet ICBM success was announced to the world five days later. This was, however, overshadowed by what was to come. As planned, the military success of the R-7 was paralleled by more peaceful developments. On 4 October, Sputnik 1 (the name of the satellite in the nose-cone section) was pitched into space by an R-7 rocket and began an elliptical orbit of the Earth (taking just 93 minutes) so that it could be seen across the planet even though it was only 23 inches in diameter. Sputnik 1's course took it 588 miles from the Earth at its furthest point and 141 miles at its nearest. It continued like this for ninety-two days, until it was drawn back into the Earth's atmosphere at which point it disintegrated as expected. All this was initially met with little fanfare in the USSR as it was seen as a small, if significant, step towards much greater things. However, in the US and the West, it caused a sensation. Ordinary Americans could see the progress of the Soviet moon with their own eyes, and pick up the simple 'bleep bleep' of its transmitter on their radios. While some marvelled at this sign of human progress, others, including prominent figures, were chilled at the meaning of this

First Steps to the Stars: Sputniks and Canine Scouts, 1956–1960

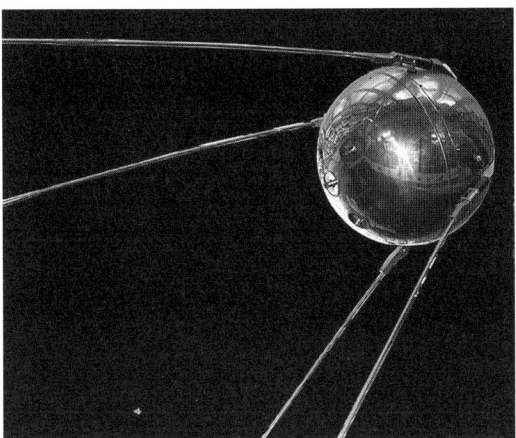

Model of the Sputnik 1 satellite. (*NASA via Wikimedia Commons*)

leap in technology from a country whose engineering capability was thought to be limited to tractors. Had the Americans studied what was being openly discussed in the USSR press, they might have been forewarned. On this occasion, they failed to separate truthful accounts from the common output of dreamy propaganda. An article before the launch by the journalist Riabichov in the popular magazine *Iskra* (Spark) had accurately described Sputnik 1's planned orbit of the Earth. Later, the US expert and writer Martin Caidin (1961) enumerated some thirteen official statements between 1955 and Sputnik 1's launch that accurately predicted facts surrounding the plans, but the idea of Soviet secrecy and double-dealing in the Year of

Soviet postcard of Sputnik flight, 1957. (*Author's collection*)

Soviet postcard celebrating Sputnik successes, October 1959. (*Author's collection*)

Children listen in to Sputnik 1. (*Novosti Soviet Sputniks, 1958*)

International Geophysics persisted. Eisenhower, the president, dismissed the news as little more than a stunt, but this was not shared by the military establishment.

The US immediately stepped up their efforts to keep up with, and overtake, their adversaries in what had suddenly and dramatically become a space race – NASA was formed the following year, in 1958. Meanwhile, congratulatory telegrams poured into Moscow and Soviet embassies throughout the world. For the Soviet scientists behind the project, Sputnik 1 seemed only a small part of their highly secret and well-hidden plans, but to ordinary people everywhere it represented a moment of huge significance in global history. The Soviet establishment did not take long to realize this and the machinery of carefully manicured information and propaganda soon ground into operation.

Dogs, Monkeys and Other Life Forms

From the late 1940s, the Soviet space team had been looking ahead to manned space flight. Here there was much to consider beyond engineering regarding the means of achieving such a possibility. How would a human being cope with the forces that the body would have to withstand: weightlessness, g-forces, psychological issues and countless other matters? Korolev appointed a specialist veterinary surgeon, Vladimir Yazdovsky, to head the biological programme in the Institute of Aviation Medicine as early as 1948. While the Americans, when they reached this stage, focused on primates (monkeys, and later chimpanzees) for their tests and experiments, the Soviets turned to canines. Dogs had long featured in the Soviet imagination in connection with science and human progress. During the Second World War a trained mine-hunting dog, *Dzhulbars*, had become famous and won the favourable attention of Stalin himself. The Russian physiologist, Ivan Pavlov, had studied conditioning by experimenting on dogs in the late nineteenth century. Dogs were popular in the USSR as both working animals and pets and their choice to pioneer a trail for humans seemed natural and chimed with the public imagination. Yazdovsky's expertise was with dogs and they were chosen based on their known physiology, aptitude for training, communication skills and sociability with people. Dogs were also in plentiful supply and the hardiest and most resilient ones could be found among the strays living on the streets of Moscow – which lent an additional affinity with ordinary working life and people.

Dogs were first sent into space in Soviet rockets in 1951. The first two, Desik and Tsygan (Gypsy – a common Soviet dog name) survived the first vertical ascent of 62 miles in July 1951. Their sealed air-supplied capsules successfully parachuted back to the steppe, from which they emerged happily. Another five launches took place for similar experiments on the effects of weightlessness, temperature, possible radiation, vibration and noise, and vacuum. Three were successful but the dogs died in the others, including a landing where four dogs were killed when their capsule parachute failed to open. Further tests took place with a series of nine launches in 1954. For these, the dogs were fitted with individual space suits that supplied air and met other life-support and sanitation needs. Five out of the twelve dogs did not survive – but much was learned. In 1957, with the R-7 rocket nearing readiness, a further series of flights took place where the dogs were located inside a sealed cabin, designed to detach from the nose cone once in orbit. The dogs were paired and rose to an altitude of 120 miles – twice what had previously been achieved. By this time the dogs had been separated into groups according to their temperament – all were female as this made it easier to manage bodily functions. All also underwent surgical procedures to aid monitoring of their health. Naturally, the dogs all formed attachments to the team members involved in their training and preparation.

As the launch of Sputnik 1 neared, the next stage of a programme intended to progress to manned space flight, the Soviets put out carefully contrived publicity about their plans. An early statement following Sputnik 1, in the newspaper *Pravda*, read:

> In order to make the transition to manned space flights it is necessary to study the effect of space-flight conditions on living organisms. To begin with, animals will be used for such studies. As was done with the high-altitude rockets, the Soviet Union will launch a Sputnik carrying animals as passengers. Detailed observations will be made of their behaviour and their physiological processes.
> (Quoted in Riabchikov 1971, p148)

There was also publicity about the dogs which would strike a chord with ordinary people in the USSR and trumpet Soviet achievement globally. While technical detail and imagery were carefully avoided, some admissions were made, including the fact that dogs had reached an altitude of 62 miles, or 100 kilometres (actually achieved many years earlier and already surpassed). The three dogs, all flight veterans, were presented for press and public display in their 'space suits' – coats whose real purpose was to hide the surgically fitted wires protruding from their bodies. All were well socialized and affectionate. They were lauded as canine scouts, pointing the way for human space exploration.

Sputnik 2 came fast on the heels of Sputnik 1's success. On 3 November 1957, Laika the dog was launched into space in a nose cone similar to that of Sputnik 1, at the head of an R-7 rocket. The date was chosen to coincide with the 40th anniversary of the October 1917 Bolshevik Revolution and built on the worldwide acclaim given to Sputnik 1. Khrushchev, at this point, was at the peak of his popularity as a Soviet leader and was able to point to real Soviet achievement. Alas, Laika's glory was short lived: she was never intended to be returned to Earth as the technology had not yet reached the stage required. After four orbits (the exact time is still the subject of differing reports), she died from overheating after the air

First Steps to the Stars: Sputniks and Canine Scouts, 1956–1960

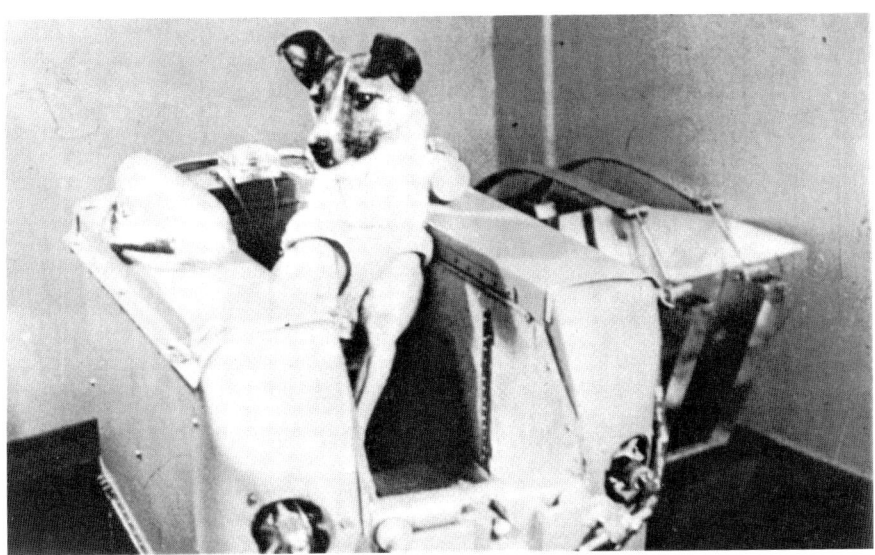

Laika, the dog in Sputnik 2. (*Novosti* Soviet Sputniks, *1958*)

conditioning system malfunctioned. After 162 days in space and 2,500 orbits, Sputnik 2, as projected, re-entered the atmosphere and broke up, with debris landing deep in the Amazon jungle of South America. Data transmitted early in the return flight provided important information, including proof that life could be sustained in space orbit.

Korolev's ultimate goal was interplanetary space travel; thus another facet of the Soviet space programme from this early stage concerned exploratory missions to the moon (a theme that will be picked up in Chapter 7). Shortly after Sputnik 2's 'successful' mission, Korolev wrote enticingly in *Pravda* (as Professor K. Sergeev – the name he always used to hide his real identity):

> The time will come when a spacecraft carrying human beings will leave the Earth and set out on a voyage to distant planets – to remote worlds. Today this may seem an enticing fantasy, but such in fact is not the case. The launching of the first two Soviet Sputniks has already thrown a sturdy bridge from the Earth into space, and the way to the stars is open. (Quoted in Riabchikov 1971, p151)

Such dreams were lost on some: Laika's death prompted anger from canine and animal rights' groups internationally (especially in the UK) and even some criticism within the USSR. However, she was certainly not the first life form to be sent to a high altitude in a rocket. That credit belongs to some fruit flies that were launched by the Americans in a German V-2 rocket in New Mexico in February 1947 – surviving their 68-mile-high journey and being recovered from their capsule that parachuted back to Earth. On 14 June 1949, Albert, a Rhesus monkey, reached an altitude of about 83 miles but died on his return to Earth when his capsule's parachute failed. The same fate befell a mouse launched by the Americans in 1950. These experimental US flights continued throughout that decade, many involving high-altitude balloons, with an attrition rate among the monkeys aboard of about two-thirds. US scientists favoured monkeys because of their biological similarity to humans. In December 1958, just over a year after Laika's flight, the Americans sent Gordo, a Squirrel monkey into space aboard a Jupiter rocket. He was lost when his parachute failed to open but proved that re-entry to the Earth's atmosphere at 10,000mph could be survived. On 28 May 1959, the Americans successfully recovered a Rhesus monkey named Able and a Squirrel monkey named Baker from an altitude of 360 miles and a flight of 1,700 miles. Although Able died of an infection soon afterwards, Baker went on to have a long and reportedly happy life in the US Space Centre in Huntsville, Alabama.

1959 saw a further series of bizarre launches of animals into space as research in both the USSR and US focused on being the first to get a human into space. This included a Soviet rabbit, Marfusha, who made a trip with two dogs in July, and less successfully, two frogs and twelve mice whose American launch in September ended when their rocket was destroyed at the launch. The 1960s began with successive launches of more monkeys and a chimpanzee (USA) and mice, frogs and a guinea pig (USSR). By the end of the decade all sorts of animals and life forms including amoebae, bacteria and plants were making the journey into space aboard French and Argentinian, as well as American and Soviet flights. Between the 1950s and 1960s the Soviets launched fifty-seven passenger journeys with dogs, some making more than one journey. The first deep space flight by animals was also a Soviet first; two tortoises and various other life forms including mealworms and wine flies circumnavigated the moon in September 1968 and were successfully recovered – all suffering some weight loss but otherwise unharmed.

The major event of 1960 was the launch of Korabl-Sputnik 2 (often referred to as Sputnik 5) which carried the dogs, Belka and Strelka, a rabbit, forty mice, two rats, fruit flies, various plants, strips of human skin and other specimens. Importantly it also carried two television cameras able to transmit moving images back to Earth. The flight on 19 August followed an unsuccessful similar attempt that had blown up at the launch stage the previous month. Mistakes were traced back to the rocket assembly plant and rectified for the new mission, the aim of which was to orbit the Earth and return the live cargoes home safely. News of the successful launch was ceremoniously relayed to a courtroom, where the trial of shot-down American U-2 spy plane pilot, Gary Powers, was concluding with his conviction: a double message for the world's media about Soviet technical superiority. During the Sputnik mission, the dogs were closely monitored to find out the effects of weightlessness and the results were not especially positive: both were seen to be stressed and uncomfortable. After twenty-seven orbits of the Earth, the capsule re-entered the Earth's atmosphere and successfully landed. Both Belka and Strelka were found to have recovered quickly and were in good condition.

Soviet postcard to celebrate Sputnik 3 and its passengers, Belka and Strelka, 1960. (*Author's collection*)

* * *

Although the Americans had beaten the USSR by recovering a space capsule, Discoverer 13, from a flight a few days before Sputnik 5, the Soviets had achieved this with life forms aboard and were again basking in the glory of another significant first. Belka and Strelka achieved great fame and were shown around the country to admiring audiences,

72 Soviets in Space

SPUTNIKS IN SOVIET CARTOONS

"To what do you attribute your scientific success," asks the press on the right. And the young Soviet technician answers, "To forty years of Socialism".

English-language cartoon from 1958. (*Novosti* Soviet Sputniks)

including numerous schoolchildren. Belka went on to have puppies, one of which, Pushinka (Fluffy) was presented to Caroline Kennedy, daughter of the US president. Other dog space flights followed over the course of the next few months – some involving dummy astronauts that were successfully ejected from the capsule before landing – with the dogs making it safely back to Earth. All these pointed towards one of the most significant events in human history that would take place in April 1961.

Chapter 5

Workers in Space: The First Cosmonauts, 1961–1964

Everything was unusual, full of colour. Space awaits not only the scientists who can see everything for themselves, explain and analyse it. It awaits its own poets and painters also … Each of the oceans and seas, like the continents, had its own colour and characteristics. The very first turns took me over the Pacific and Atlantic oceans and though from that height they looked completely calm, under three and five power horizon scanners, I could distinguish the endless rollers lazily chasing each other across the Pacific and the long swaths of the Atlantic. The Indian ocean was indigo blue, the Gulf of Mexico salad green and the central part of the Mediterranean gleamed like an emerald.

Gherman Titov (in Burchett & Purdy 1962, p126)

Konstantin Tsiolkovsky is credited with predicting that humans would achieve space flight no earlier than 2017. From the isolation of his log cabin in Kaluga, he could not have foreseen the rapid advances in technology and engineering spurred by war and developed in its aftermath. He was out by a mere fifty-six years – a trifle when one considers the magnitude of such an achievement, almost the same number of years after the first powered aircraft flight in 1903. The accomplishment of chief designer Korolev and his team, in succeeding with the space orbit around the Earth of Yuri Gagarin on 12 April 1961, was truly remarkable, and, as we shall see, was used to project an image of Soviet superiority in science, technology and human progress. By 1960, the Soviets remained in the lead despite American attempts to retain prestige. The worthy flights of the X-15 experimental aircraft 60 miles up to the edge of space were presented in one publication as actual space flight, although detailed reading revealed the truth (Caidin 1959). The race, however, was close, and each side maintained a wary watch on the other. All this had a profound impact on the Soviet people.

Nikolai Kamanin and the Training of the First Cosmonauts

In keeping with the Soviet habit of making broad advance statements of progress but detailing achievement only after the event in bulletins – failures covered up in their entirety – the selection and training of the first cosmonauts was undertaken in secrecy. Those chosen were initially unaware of the ultimate goal for which they were being prepared. Issues of dealing with weightlessness, velocity and a multitude of other technical and physical issues could apply to very high-altitude flying as well as space travel. Preparation also had to take into account the role success would bring to these young people – they had to be ambassadors for the Soviet state and represent communist aspiration as well as present accomplishment. While Korolev's team dealt with the technical issues alongside all the other secret aspects of the space programme, air force general Nikolai Kamanin (1908–1982) was chosen to manage the vital psychological preparation and overseeing of the cosmonauts' private lives.

Kamanin was a good choice for this role, which he assumed as deputy head of Training for Spaceflight in 1960, effectively putting him in charge of the training centre and representing the air force, from which candidates were drawn, in all matters. There are parallels between Soviet heroes of the Stalinist 1930s and the cosmonauts; all were presented in their respective eras as representing Soviet ideals about the pioneering but collective spirit of socialism and communism. This contrasts with the individualist of popular American folklore typified by the popular TV, movie and comic cowboy heroes of that era. Flyers were especially revered in the USSR in the 1930s and the names of those who set records or engaged in daring missions were well known. Kamanin represented this tradition; in 1934 as a young pilot he had been involved in the air rescue of the crew of a ship on an Arctic exploration expedition – the *Cheliuskin* had been crushed in the ice. For this, he and a fellow rescuer were made the very first subjects of HSU status. He had gone on to be a successful and decorated air force commander in the war – in which his son Arkady fought in one of his units as a 14-year-old military pilot. Kamanin knew about the uncertainties the cosmonauts faced, and also the pressures of fame that would come to those who survived. He was a hard taskmaster, brooked no disobedience and exercised absolute control over the lives of those he helped select for this very special task – using his

Kamanin (back row centre) and cosmonauts in Moscow, 1 December 1969. (*A. Makletsov RIAN via Wikimedia Commons*)

power sympathetically and winning the respect of those in his charge. He got to know these young people and their families very well and used this knowledge to enhance both training and the public profile that would follow upon success.

The pioneer rocket scientist, Tikhonravov, is said to have coined the term 'cosmonaut' – a different Soviet construct from the common English term 'astronaut', popularly used to describe the American spacemen of the same period. One of the early cosmonauts, Leonov, recalls in his memoirs that he and his colleagues were proud of being cosmonauts as this modestly described their task of voyaging within the solar system. They considered the American term boastful as it implied flying between stars (Scott & Leonov 2005).

The selection process began in 1959. The specifications for nominees decreed by Korolev were: men aged 25 to 30, height 5ft 6in to 5ft 7in, weight 340–351 pounds – all to fit within the tiny capsule. In terms of specialist knowledge, automation was such that cosmonauts were initially only expected to carry out particular, and not necessarily regular programmed, tasks. Thus, engineering and scientific skills were unnecessary (this differed from the American programme where more flight control functions were expected of potential astronauts). Air force fighter pilots were an obvious choice for potential cosmonauts as they were already trained to a high level and understood aeronautics. Also, they were familiar with military discipline, and enough could be found to meet Korolev's physical specifications. Records of some 3,000 pilots were scrutinized and most discarded for reasons of height, weight and medical history. Those left were interviewed on 3 September 1959, and were told only that this was for top-secret special flights – some caught on quickly when certain questions were asked. The interviews also focused on attitudes, quality of life, personal goals and mood. Two hundred went forward to the next stage of selection – groups of twenty at a time were examined in the Central Scientific-Research Aviation Hospital in Moscow. The majority failed the rigorous physical tests and eventually, at the end of the year, twenty emerged as potential cosmonauts. This was a higher number than envisaged due to Korolev's insistence that a group larger than that was being similarly prepared in the USA (seven). Kamanin's appointment soon followed and he was allowed to scrutinize the candidates before final approval was given for their appointment in February 1960.

The successful nominees were brought together for their training at specially converted premises at the I.V. Frunze airfield in Moscow, arriving in March 1960. The names of these were only released in full some twenty-five years later – otherwise, their identities only became known on their return from successful space missions. The twenty were Senior Lieutenant Ivan N. Anikeyev (27), Major Pavel I. Belyayev (34), Senior Lieutenant Valentin V. Bondarenko (23), Senior Lieutenant Valery F. Bykovsky (25), Senior Lieutenant Valentin I. Filatyev (30), Senior Lieutenant Yuriy A. Gagarin (25), Senior Lieutenant Viktor V. Gorbatko (25), Captain Anatoliy Y. Kartashov (27), Senior Lieutenant Yevgeniy V. Khrunov (26), Captain Engineer Vladimir M. Komarov

(32), Lieutenant Aleksey A. Leonov (25), Senior Lieutenant Grigoriy G. Nelyubov (25), Senior Lieutenant Andrian G. Nikolayev (30), Captain Pavel R. Popovich (29, Senior Lieutenant Mars Z. Rafikov (26), Senior Lieutenant Georgiy S. Shonin (24), Senior Lieutenant German S. Titov (24), Senior Lieutenant Valentin S. Varlamov (25), Senior Lieutenant Boris V. Volynov (25) and Senior Lieutenant Dmitriy A. Zaykin (27). Several fell outside the original age criteria but were admitted due to their outstanding performance during the selection tests and, in the case of Komarov and Belyayev, their great experience as pilots.

The training, which involved endurance, callisthenics and other physical tests was accompanied by exercises including weightlessness and arduous use of a centrifugal force simulator. Lectures were also given in communication, space technology, astronomy, navigation, geophysics and space medicine. The would-be cosmonauts were also subject to psychological testing, including prolonged isolation and confinement in small spaces. Parachute jumps to simulate the skills required for safe landings in all environments and temperature were also conducted: each completed some fifty jumps. Towards the end of training a simulated space capsule was introduced to the programme. After one year, and a series of exams, the trainees passed out officially as cosmonauts. In late May 1960, a group of six was chosen for continued accelerated training as all twenty could not be offered this experience at the same time. The Vanguard Six were Gagarin, Kartashov, Nikolayev, Popovich, Titov and Varlamov – one of them would be the first man in space. Korolev met with this group for the first time in June, and in July they were shown the spacecraft at the OKB-1 production facility. At the end of June 1960, cosmonaut training was moved to a new purpose-built centre outside Moscow named Zelenyy (green) which was close to OKB-1 and the main Monino airfield. Today this is better known as Zvezdniy Gorodok (Star City) where cosmonauts continue to live and undergo training.

By July 1960 the Vanguard Six had been reduced for medical reasons – Kartashov and Varlamov were quickly replaced by Nelyubov and Bykovsky. By this time plans for the spaceship, named Vostok 1 (East 1) were advanced. The pressure was on because the Americans had announced that they expected to launch a man into sub-orbital spaceflight in January 1961. Korolev was determined that this should be overtaken by December 1960 with a manned space flight that would orbit the Earth.

This became official policy. Gagarin later recalled the training regime: while the Soviets were confident that they were exceeding the Americans with their rocket-fuel system that could propel a greater weight into space – and therefore more equipment to the extent that the pilot represented only 2 per cent of the payload, the human factor was still an uncertainty. He and the others were subjected to extreme tests to weigh their ability to withstand the crucial factors of take-off, acceleration, orbiting, re-entry and deceleration, and landing. All these also involved temperature extremes, loneliness, silence, fear, idleness and absence of communication. Off-duty hours were spent sleeping and reading, with only the occasional beer and a total ban on smoking. Early on in the training, the express lift of the twenty-eight-storey building at Moscow University was used for weightlessness tests. This involved releasing the lift, with its human occupant, to plummet 500 feet before sudden arrest at the bottom by special air brakes. The passenger had his feet on the floor for less than a second before becoming weightless and being suspended between floor and ceiling – a terrifying but cost-effective way of achieving zero gravity. Gagarin also spent hours on a vibration table – being shaken 200 times a minute, and was also subjected to oxygen starvation experiments. He was given tasks to complete while gradually and unknowingly losing consciousness – all with the real goal of testing endurance. Leonov later recalled that the centrifugal-force simulator rotated the trainees round and round at terrific speed until they lost consciousness, all monitored closely by the watching doctors and scientists.

The rewards for this hardworking, elite, but hidden group were spartan compared with their American counterparts, reflecting Soviet values and ideals. Their salary amounted to the equivalent of $100 a month at a time when the first group of highly publicized American astronaut trainees had received half a million dollars for their personal stories from *Life* magazine. When asked about the matter of salary by Western journalists after his success in 1961, Gagarin parried this by advising that a cosmonaut's salary was commensurate with his needs and they were all quite happy as citizens of a state which looked after all its people. The cosmonauts did enjoy some privileges, with accommodation (including for families) in Zelenyy that was eventually built to their collectively agreed specification – two interconnected blocks of flats with communal areas for meetings and relaxation.

Gagarin emerged successfully and unscathed from his training, as did his colleague Gherman Titov. Kamanin now had to make the difficult choice about who would be the first human in space. His diaries reveal that Gagarin was chosen not because of his superior abilities, but because Titov was even better (Kamanin 1995–1997). The first flight was planned to be almost entirely automated, while the second, which would follow quickly, would involve more duties and effort on the part of the pilot – and Titov was considered the better of the two in this respect.

Meanwhile, the success of the Vostok mission involving the dogs Strelka and Belka, and their living companions, in August 1960, as well as other test flights, all pointed to the possibility of a December date – albeit with some concerns about the haste of the plan. However, disaster befell the Tyuratam launch site when an ICBM missile, designed by another team headed by Mikhail Yangel, exploded on launch on 24 October 1960 – an accident that killed some 130 personnel, including many leading scientists and engineers. Although this programme was not directly related to the Vostok space programme, it caused an inevitable delay. A new launch date of February was agreed – it was still possible with this to outpace the American programme which had also suffered setbacks and delays. Further delays were resulting from equipment malfunctions discovered in other Vostok test flights during this period. One of these, unfortunately with dogs on board, was deliberately scuttled in early December when a malfunction made a landing in foreign territory a possibility. Such was the obsession with secrecy that this could not be countenanced. A further flight with dogs later in December also malfunctioned and was aborted soon after launch; this time the passengers survived and were successfully recovered from a remote and snowbound area of Siberia. With two unsuccessful Vostok missions in a row, the February plan for the piloted flight was abandoned. Meanwhile in the US, the schedule for the manned sub-orbital flight was now focused on a May launch. Everything now rested on two successive automated Vostok test flights in March, designed to simulate the manned flight as much as possible. These were successful and plans were advanced for an April manned flight that could be celebrated on May Day. Unusually, given Soviet secrecy, Khrushchev announced in a press interview in mid-March that a manned space flight was imminent, which was widely reported. However just a few days before the second of the March test flights, tragedy befell the Soviet space programme.

The Short Life and Tragic Death of Valentin Bondarenko

The youngest of the initial intake of cosmonaut trainees disappeared from the programme, and temporarily from history when he became the first human casualty of the programme on 23 March 1961. Valentin Bondarenko was fairly typical of the first batch of trainees in terms of aptitude, temperament and background. Born in 1937 in the Kharkov region of Ukraine, Valentin's father was a tailor and his mother worked in the same Kharkov factory as a cutter. Valentin's father was a true war hero: wounded and captured in 1941 near Kiev, he later escaped and joined the Odukha partisan group, fighting with them until 1944. Meanwhile, at home in Kharkov, under German occupation, Valentin and his mother endured severe hardship. After the war, he attended the Air Force High School in Kharkov and, with an interest in flight, became involved in flying through DOSAAF – the paramilitary youth organization attached to the Komsomol (Young Communist) organization. He graduated from school in 1954 and went straight to the Voroshilovgrad (now Lugansk) Military Aviation School in Ukraine, and from there to the Armavir Pilot School in Grozny in 1955, graduating in 1957. In 1956 he married Galina Rykova, a medical worker, and their son was born later that year. Commissioned as a second lieutenant, Valentin served as a fighter pilot in the Baltic military region. He started his cosmonaut training on 31 May 1960 where he was nicknamed Tinkerbell and liked for his mild manner and pleasant character. Valentin had a good singing voice and was a skilled tennis player.

Cosmonaut Valentin Bondarenko with his wife and child. (*Kharkov Planetarium*)

Early cosmonaut training involved a high degree of managed risk, but new areas of human endurance, as well as scientific knowledge, were being explored and tested; it was inevitable that some lessons would be learned the hard way. On 23 March 1961, Valentin was ending the fifteenth day of an endurance test in a high-pressure chamber at the training facility in the Military Aviation Hospital in Moscow. He had just completed his day's scheduled activity, had removed the sensors from his body and was washing his skin down with an alcohol-soaked cotton ball which he then threw aside. This accidentally landed on a hot plate he was using to prepare a cup of tea and caught fire. His attempts to douse the flames in the highly volatile chamber, which was 50 per cent oxygen, only served to spread the fire to his woollen clothing. He soon became helpless as the flames engulfed his body. The horrified attending doctor spent half an hour watching before pressure could be reduced enough to open the door and attend to Valentin. By this time the fire had used up all the oxygen in the chamber and he was almost completely covered in third-degree burns. When attempts were made to treat him, the hospital specialist had to feed an intravenous drip through the soles of his feet – the only area of his body (protected by his boots) where a blood vessel could be found to insert a needle. Valentin survived only eight hours before succumbing to shock from burns that had consumed his body, apparently watched over for some of this period by Gagarin. His wife received a posthumous Red Star award, a special pension and continued to live in Zelenyy (Star City) with the couple's child. The death, however, was hushed up and not revealed publicly until the 1980s. The American crew of Apollo 1 died in a similar fire in 1967 but by this time the danger of fire in high oxygen environments was well known – they had not been in 1961. Valentin is buried in the cemetery in Lipovaya Roshcha, Kharkov, where his parents were living. His grave was only marked as that of a cosmonaut hero in the 1980s. As he was carried out of the pressure chamber, Valentin Bondarenko is alleged to have shouted that the accident was his fault and that no one else was to blame. That was not the view of Kamanin, whose diary acknowledges what was said about the cause of the fire but records concern about the testing regime.

First Man in Space

Just two days after Valentin Bondarenko's fatal accident, the final test flight, that of Korabl-Sputnik 5, was successfully launched and recovered, along with its animal and plant-life passengers. Later that month, Korolev hosted a reunion of surviving GIRD scientists and engineers with whom he had worked from as far back as the early 1930s, and although revealing no details of the imminent human launch, showed them the spacecraft destined for that voyage – a significant moment linking the past of Soviet rocket science with the future.

With all pre-testing now complete, Gagarin blasted off into space aboard Vostok 1 on 12 April 1961 – a historical moment of world importance and the greatest-ever achievement of the Soviet space programme. Twenty million horsepower of rocket engine hurtled him quickly beyond anywhere previously reached by mankind and the first stage of the rocket separated and fell away towards Siberia as planned. Soon the Earth's atmosphere was left behind and the second stage of separation left Gagarin in his tiny capsule heading at 17,500mph on man's first orbit of Earth. Short, yet calm, clear radio exchanges confirmed that all was well and the enigmatic hit tune *Moscow Nights* was played to Gagarin from Earth. Weightlessness in zero-gravity conditions demonstrated no unpleasantness. All too soon warning lights showed that the orbit was almost complete and the moment of potentially dangerous re-entry over Soviet territory was nearing. Although operations were being exercised remotely, Gagarin could have assumed control himself had there been any problem, even continuing onto a further orbit. However, this was not necessary, the retro-rockets that would reduce speed for re-entry worked perfectly and Gagarin began his rapid descent to a spot some 90 miles from the Baikonur launch site. The story differs here according to when it was told: at the time it was announced (and backed by the witness statements of the peasants who were on the scene working in the fields) that Gagarin emerged on the ground from his capsule. In fact, he ejected as planned and parachuted down separately at 1055 hours Moscow time. All this lasted just 108 minutes from the moment of launch to landing. This is the story told proudly to the world, and most of it was accurate – only in post-Soviet times would it be revealed that there were pre-launch delays due to technical failure, and tense and potentially disastrous minutes during space flight when the capsule would not separate from the second stage of the rocket.

Such was the secrecy at the time that it was impossible to convey fully to the world the detail of the flight, and the feeling of achievement felt by the members of the team led by Korolev. The 2013 Russian movie, *Gagarin:*

Workers in Space: The First Cosmonauts, 1961–1964

First in Space probably comes as close as might be possible to portray the drama and excitement of those 108 minutes. While Gagarin's role as an outstanding leader is possibly overblown, the bravery and character of the man seem accurate when set alongside what else we know about him. The film also presents a reasonable portrayal of the technology involved – and lends insight denied at the time through the limits of film technique and secrecy.

Yuri Gagarin just before lift-off, 12 April 1961. (Космонавмы рассказываюм Издательство *1964*)

Gagarin became an immediate and recognizable hero across the world. Not even Kamanin, who made every effort to exercise control of events and public staging, could have foreseen the impact of the still-anonymous Korolev's achievement. Gagarin was soon on a flight to Moscow from Kazakhstan, escorted by his former colleagues in MiG fighter jets. His landing was followed by a ceremonial and symbolic red carpet walk from the aircraft to meet Khrushchev and other dignitaries on the podium in front of the world's press and television. Unbeknown to those watching – but a much-discussed matter since – was the irritation and tension felt by Gagarin personally as one of the garter straps holding up his sock came loose and threatened to trip him up in front of an audience of millions on this most important day of his life. The moment passed, Khrushchev made the first of many speeches that day and the party left for a reception in Red Square by motorcade.

As usually happened on important occasions, workers in the largest Moscow factories and workplaces were given time off to line the streets and they now made their way to their allotted stations on Gagarin's route. This, however, was different: the spontaneous joy in the air was more like that of the victory celebration on 9 May 1945. Hundreds of thousands, many more than expected and arranged, thronged the route to applaud their new hero cosmonaut. This was a scene repeated wherever

Gagarin's first celebrity steps, Moscow Airport, April 1961. (Утро Космической Зры, *1961*)

Moscow crowds welcome Gagarin. (Утро Космической Зры, *1961*)

Gagarin went from then on, with stage-managed tours of the Soviet Union, neighbouring socialist-friendly countries, and the West. It would eventually take its toll.

Gagarin's flight indelibly marked Soviet citizens and would be remembered with pride for many years to come. Alexander Kudinov was a 13-year-old school student living with his parents in a village in the Kursk region. He readily recalls the excitement on learning over the radio of the successful orbit of the Earth, with a national holiday taking place almost spontaneously. Everyone was caught up in the enthusiasm and pride in beating the Americans. He and his friends all wanted to be cosmonauts. By that time Alexander was attending school in the nearest town 19 miles from his home, bicycling there every day in the summer and staying over in a hostel in the winter. This contrasts favourably with the circumstances of his village education at the time of the first Sputnik flight in 1957. Alexander's village had no electricity until 1962 and primary schooling was in the huts of various villagers, in space shared with animals when it was really cold. This was the reality of life for many millions of rural citizens.

Workers in Space: The First Cosmonauts, 1961–1964

English-speaking newspapers from across the world mark the first human in space. (Утро Космической Эры, *1961*)

Not all Soviet citizens were as interested in the space programme. Veniamin Nikitsky, born in 1937, and living in Kharkov had learned about the possibility of space flights in 1956 from a prohibited *Voice of America* broadcast. Veniamin listened in on a crystal radio set, managing to hear these broadcasts even though they were jammed by the government. He discussed space flight with his uncle who was an engineer in a large factory but was told that this was not possible – the Soviets kept their plans very secret until success was assured. After the Sputnik 1 launch in 1957 and the official announcements, he and his friends tried to track it through the sky. By the time of Gagarin's flight in 1961, he had become disenchanted with the Soviet state and took little interest in the announcement and celebrations.

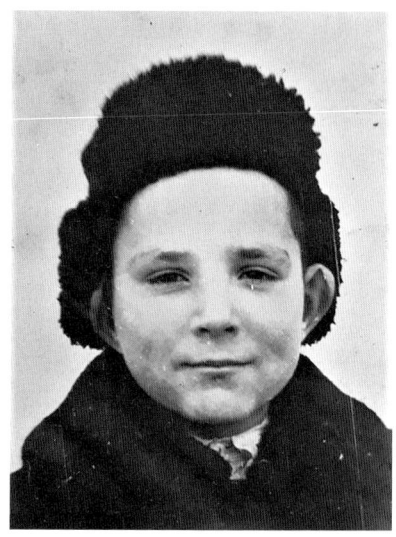

Alexander Kudinov, 1962. (*Kudinov family*)

Generally, life was relatively comfortable in the cities and people were more in tune with the technological advance represented by Gagarin's flight. Russia's TV12 channel featured reminiscences on the occasion of the 50th anniversary in April 2011 and several participants recalled the awakening of interest in science and technology, with an influx of highly motivated students into higher education institutes in the cities. All recalled the sheer joy on the streets as Gagarin's cavalcade made its journey to Red Square. People hugged and celebrated together, all hoping to catch a glimpse of this young people's hero.

Veniamin Nikitsky, 1955. (*Nikitsky family*)

Yuri Gagarin: The Man

Gagarin fitted his role perfectly – his background and personality were the very model of the type of man who would build communist society. He was charming, with an almost permanently fixed smile, and people took to him readily – making him a truly popular figure with whomever he met – including Western dignitaries. This of course was no accident and was just one of the criteria used to choose the early cosmonauts. In July 1961, he met the 35-year-old Queen Elizabeth for a hastily arranged meal in Buckingham Palace. Despite all his communist beliefs, Gagarin was said to be overawed by the presence of a real royal – a Russian biographer (Danilkin 2011) reports that as he sat next to her, he touched her leg under the table to make sure she was real. The Queen was charmed and helped him choose the right cutlery; Gagarin commented that perhaps she was mixing him up with an officer of her own Royal Air Force. He was a great asset to Soviet diplomacy.

So, who was this man? Yuri Gagarin was born in a small town near Gzhatsk (renamed Gagarin in 1968), Smolensk province, Russia, on 9 March 1934. He was one of four children. His parents worked on a collective farm: his father as a storeman and his mother (who was actually from St. Petersburg and quite well educated) with the dairy herd. More humble, hard-working origins could not have been invented. Life was happy for Yuri until the war came in 1941 when the area of their home was occupied after fierce battles in the immediate neighbourhood. Yuri changed from being a carefree child to a serious one very quickly and the family suffered years of intense privation and hardship. The Gagarins were ejected from their home by occupying German troops and had to live in a dugout in the ground. Yuri's brother Valentin was hung from a tree by a particularly sadistic soldier who had taken a dislike to him; fortunately, the assailant was distracted by colleagues and no permanent harm was suffered. But later both Valentin and his older sister Zoya were taken to Poland to work as slave labourers. The family suffered further indignities, including an assault on his mother, Anna, by a German soldier which left her with a gash in her leg from a scythe. During this time Yuri witnessed aerial battles and became fascinated with flight. After the war, and thankfully reunited as a family, the Gagarins moved to Gzhatsk where they built a new home from the wood recovered from the wreckage of their old one. The younger children, including Yuri, started school, where he was mesmerized by a teacher who had served in the air force. Yuri applied himself well and was especially good at maths and physics.

In 1950, having finished schooling and now in need of a trade, Yuri was enrolled at a technical school in Moscow and served an apprenticeship as a steel foundry worker. He did well and after a year went on to the city of Saratov for further technical education – mostly associated with tractor design and maintenance. It was here that he was able to join the DOSAAF flying club through his Komsomol membership, where he learned to fly in his spare time. Yuri also joined various sporting clubs to build up his physical fitness, playing

A young Yuri Gagarin admires the statue dedicated to pioneer woman flyers, Saratov, 1955. (Юрий Гагарин На Земле Саратовской, *1972*)

volleyball and basketball and impressing the girls with his water-skiing on the Volga River. He made his first solo flight in 1955 – the year of his graduation. With his keen interest and aptitude for flying, Yuri went on to the Air Force Pilots School at Orenburg, against the wishes of his parents who did not want a military career for their son. In November 1957, Yuri graduated as a fighter pilot and at the same time married Valentina (Valya), a medical assistant who worked at the Orenburg base whom he had met sometime earlier at a dance. Soon the couple was off to his first posting as a fighter pilot on a base in Zapolyarny, Murmansk, in the far north of Russia. The couple's first child, Lena, was born there in April 1959. Soviet accounts from the 1960s hold that Yuri and Valya were already discussing the prospect of space travel after the first Sputnik success of 1957. This seems unlikely – Valya married an ordinary military pilot expecting to follow him around various bases as his career progressed. The idea that he sought out the opportunity for cosmonaut training has become part of the Gagarin myth – he and the others were selected and gradually shortlisted without their initial knowledge. Family life was to be turned upside down within a few short years, in the meantime, Yuri led the life of an ordinary air force pilot. As a flyer, he was not perfect and became known for bad landings. Improvement come when he started using a cushion in the cockpit to make up for his lack of stature – 5ft 2in, which of course made him ideal for fitting into a Vostok space capsule.

After Yuri's acceptance onto the space programme, the couple moved into an apartment at Zelenyy. Come April 1961, Yuri's sudden emergence as a cosmonaut hero of worldwide importance saw journalists visit the flat and pore over the details of the couple's past, and life together – all officially managed by Kamanin and other Soviet officials. The photographs from that time clearly show the emotion and tensions involved for Valya, who cannot have been prepared for the fame that suddenly overtook her life. Over the next few years, she travelled the world and met famous and powerful people with her husband. Valya, however, did her best to avoid the limelight and preferred life at home with their two daughters (their second born in 1961), working as a medical assistant in the Yelenyy Medical Centre.

In his public speeches Gagarin promoted Soviet values and eschewed materialism, but he was quite taken with some aspects of decadent capitalism. Although presented with the latest Volga car in the USSR (which now rests on a covered plinth in his home city, Gzhatsk – renamed Gagarin), he preferred the French Matra Djet sports car given to him by the French government and used it regularly in the USSR. He was also prone to succumbing to the attentions of admiring women, including the Italian film actress Gina Lollobrigida who is said to have evaded security to turn up in his hotel room after meeting him in Moscow at the time of an international film festival in 1962. Parties in Yelenyy (Star City)

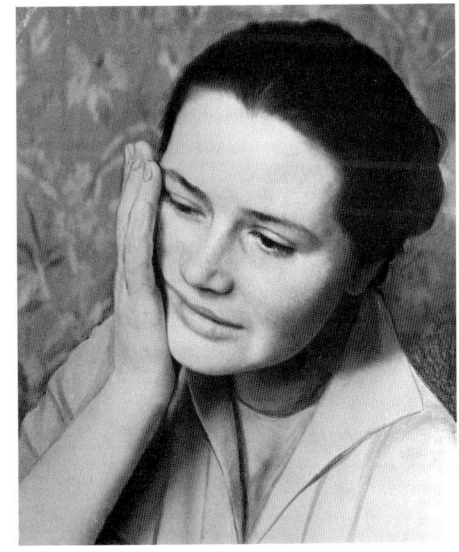

Valya Gagarin awaits news of her husband, 12 April 1961. (*Sputnik/Alamy Stock*)

Yuri Gagarin and Gina Lollobrigida, Moscow, 1962. (*TASS/Alamy Stock*)

became notorious – everyone wanted to be in the company of Gagarin and his colleagues. These would involve escapades in the early hours including impromptu water-skiing trips and hair-raising drives in his sports car. Yuri himself found the attention he received irresistible and enjoyed the partying and the alcohol.

Yuri may not always have deserved it, but in keeping with the traditional loyalty of Russian wives, Valya stood beside him and supported him as best she could. Kamanin records one episode in October 1961 when Yuri, Valya and several other cosmonauts were staying in a hotel in the Crimea with Kamanin in attendance. After a night of drinking and playing card games and chess, Yuri ordered Valya to bed and then went off in search of sex with a nurse who was quite innocently reading on her bed when he burst into her room. Valya, in hot pursuit, was banging on the door and shouting so Yuri sought escape from the situation by jumping from the balcony, falling onto a cement curb. When Valya found him, he was covered in blood with a serious head wound above his eye. This required an immediate operation and several weeks in hospital. When Yuri emerged for a star appearance at the 22nd Party Congress, he was heavily made up and had a prosthetic eyebrow. Official stories of an accident while playing with his daughter, in which he played a heroic role saving her from a fall, did not stop the rumours. Khrushchev was very angry but ordered him to attend the congress once he had checked photographs of Yuri's face. Despite a typically communist self-criticism meeting with his cosmonaut comrades, when he and Titov (who was also by now a hero) admitted excessive drinking and mistakes in their lifestyle, this all continued for many years. The pressure on this very human young man was incalculable, it all took its toll and within a few years, Gagarin was visibly overweight and pasty-faced.

The Flight of Gherman Titov

Within a few days of Gagarin's flight, on 5 May 1961, Alan Shepard became the first American in space with a flight aboard the Mercury spacecraft beyond the Earth's atmosphere and immediately back down – a considerable achievement but one of far less technical significance than the orbit already achieved by the Soviets. The USA then followed this up with a similar flight involving Virgil Grissom on 21 July. Korolev's plans were now focused on a longer mission, involving multiple orbits during which the pilot would take control. Gherman Titov, who had been the back-up for the first flight, was chosen for this because of his technical aptitude.

Titov, born in 1935, was from the Altai region of Western Siberia some 2,000 miles from Moscow. His ancestors were Russian Slavic pioneers who had settled remote, and previously uninhabited and inhospitable, areas of the Tsarist empire. His first name is taken from a Pushkin character and is unusual in Russia. The hard-working and very poor Altai peasants were enthusiastic supporters of the Bolshevik Revolution that promised them 'land and bread' and formed communes where everything was shared. In the early 1930s, these were brought together in larger collective farms. In the village of Polkovnikovo, Gherman's father, who was a well-read and self-educated shepherd, became the school teacher. The home was far to the east of the war but suffering at that time was widespread, not just through losses of menfolk, but food was also very short. On one occasion in the winter of 1945, Gherman was sent to collect a large bundle of food and nearly died in his struggle through a blizzard to return the precious cargo safely home. Later, in his teens, he caught the flying bug and did sufficiently well in his final exams to head off to the flying school in the rapidly developing community of Kostanay in northern Kazakhstan. In 1957, on his 22nd birthday, Titov passed out as an air force lieutenant and, at his insistence, was posted to a fighter squadron in the Leningrad region. He married Tamara, a Ukrainian cook he met at a dance at his base, in 1958. Unlike his colleague Gagarin, Gherman was by inclination a shy individual and avoided attention and social occasions – later, fame would change these characteristics. The following year he was picked out for special training, only hints and his broad agreement, suggesting this might be anything to do with a space programme. In training he and the others were told to tell people they were engaged in supersonic jet testing, and it took some time for him to share the truth with his wife. Eventually, once he was in the final group of trainee cosmonauts, the couple moved to Zelenyy, where their daughter Tatiana was born in 1964.

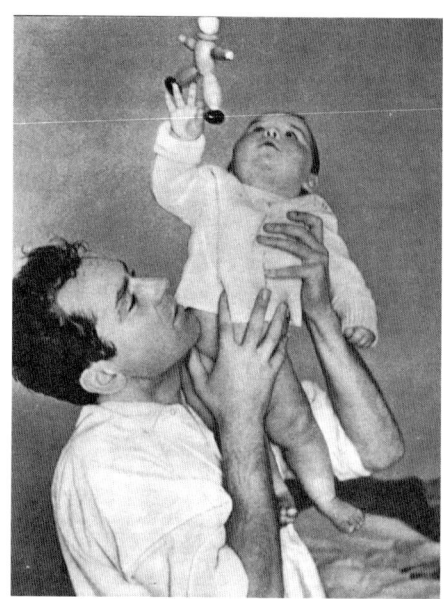

Gherman Titov and his daughter Tatiana. (Утро Космической Эры, *1961*)

Gherman Titov is still the youngest person to go into space – he was 25 at the time of his flight aboard Vostok 2 on 6 August 1961. A prime purpose of this mission was to test the cosmonaut's ability to endure a longer period in space, including eating, drinking and sleeping. The flight lasted 25 hours 11 minutes and included seventeen orbits of the Earth, during which, on Korolev's instruction, Gherman assumed control. He also took numerous photographs with a special 'Konvas' camera. Other records were also set: after eating puréed meat and liver patés, drinking blackcurrant juice, followed by solid bread and peas, he was the first person to suffer space sickness and vomit outside the Earth's atmosphere – apparently, this did not affect his operational abilities. At the end of his flight and re-entry in the Earth's atmosphere, Gherman landed by

Workers in Space: The First Cosmonauts, 1961–1964

President John F. Kennedy, Gherman Titov and US astronaut John Glenn, Washington DC, May 1962. (*NASA via Wikimedia Commons*)

parachute after ejecting from his space capsule as they fell towards the projected landing area in the Saratov region. All was reported to have gone exceptionally well.

As Gherman emerged a very public hero, he immediately came under the same pressures as Gagarin: his life would now be subjected to crafted representation and his activities, just as they had been in space, controlled and directed by others. Like Gagarin, he soon fell prey to the wishes of his admirers, men and women, to enjoy his company. Titov too liked the fast cars, women and partying that accompanied fame. In his case, there was speculation that his behavioural changes were the result of the sickness he had suffered in space, but testing proved this not to be the case.

Valentina Tereshkova Proves Girls Can Do It Too

The Americans continued their Mercury programme aimed at manned space orbit and achieved a successful three-orbit flight with astronaut John Glenn on 20 February 1962. On 11 August of the same year, after numerous technical delays that angered Soviet leaders, concerned that the Americans had now achieved similar success, the USSR again proved its technical supremacy by launching two spacecraft, Vostok 3 and Vostok 4. These were piloted respectively by Andrian Nikolayev and Pavel Popovich, who rendezvoused in space and passed close to each other to test the future possibility of docking together. They also undertook coordinated activity in front of TV cameras, so that comparisons could be made and information built upon that had been learned from Titov's flight. The pair landed within minutes of each other on 15 August. During the celebrations and interviews afterwards, Nikolayev announced that he would like to get married but had at that point no 'special' girl in mind.

In the summer of 1961, Nikolai Kamanin began to seek support for the idea of including women in the space programme. This initially found no favour but after he took the idea to Khrushchev, a plan to recruit and train five women received CPSU Central Committee approval in December 1961 (Siddiqi 2000). This was a political move designed to showcase the status and achievement of Soviet women, and of course, beat the Americans. The successful candidates would have to be like Gagarin and Titov – exceptionally fit and intelligent, but also good communists with a temperament and personality that would make them worthy ambassadors for the USSR.

Korolev, who had initially opposed Kamanin's idea, preferred fighter pilots for cosmonaut training as they combined the very quick-thinking skills of pilot, navigator, systems operator and flight engineer in one person. When he was tasked to recruit women for cosmonaut training, there were no female pilots to choose from. There had been a handful of female fighter pilots during the war, some of whom had been very successful, but they were never completely approved of and no women were being trained for that role by 1945. In fact, in the Soviet Armed Forces, by the 1960s, women had long been relegated to the type of non-combat support roles found in Western armies. Kamanin would have to look elsewhere. It was natural that he turn to the organizations that had introduced thousands to the same skillsets in the 1930s – the OSOAVIAKhIM aero clubs that had produced the women flyers of the Great Patriotic War. The air force team of physicians who had been involved in selecting the initial cohort of men, contacted DOSAAF (the successor organization to OSOAVIAKhIM) aero clubs throughout the country to identify women who might be suitable. By definition, these would be Komsomol or even Communist Party members who would do their duty for the state. A master list of 400 was soon cut down to fifty-eight with significant parachute experience who met physical requirements in terms of height and weight. The list was then thoroughly examined by a small team including Gagarin, and the women themselves subjected to medical tests and interviews just as the men had experienced in 1960. Valentina Tereshkova recalled in an interview in the 1990s (Lothian 1993) that she was called to Moscow in December 1961 for medical tests and then told to report to the training centre; her friends and family believed that she had left her hometown for special parachute training. Even her mother did not find out the real purpose of her move to Moscow until news of her space flight emerged in June 1963. By April 1962 five women had emerged from the tests and shortlisting process to start their training: Tatiana D. Kuznetsova (20), Valentina L. Ponomareva (28), Irina B. Solovyeva (24), Valentina V. Tereshkova (24) and Zhanna D. Yerkina (22). They were from a variety of working backgrounds: a mathematician, an engineer, a teacher, a secretary stenographer and perhaps the humblest in terms of occupation, Valentina Tereshkova, a textile factory worker. Solovyeva, a university graduate, was the most experienced parachutist with 320 jumps to her credit. Tereshkova had seventy-eight and Ponomareva had ten. Ponomareva, the mathematician, had the best technical and experiential CV: 320 hours' flying time, a graduate of the Moscow Aviation Institute and time served as a scientist at the Academy of Sciences; her presence in the group of five was opposed by some including Gagarin because she was a mother, but he and others of the same traditional mindset were overruled. The trainee with the least academic attainment was Tereshkova, but she had been an active member, and indeed a leader, of her local Komsomol, an important consideration.

By all accounts, Valentina Tereshkova was, and remains today, a remarkable woman. Like the men, she fitted the profile sought by Khrushchev and the Soviet leadership. She was from a working-class family in the city of Yaroslavl. Her father had been a tractor driver and soldier killed on a wartime battlefield. She was a worker and Komsomol group secretary. Work in the Red Perekop textile factory had not always been her plan. She had wanted to go to Leningrad to be trained as a railway

Women cosmonauts. From right: Valentina Ponomareva, Irina Solovyeva and Valentina Tereshkova, 1963. (*Everett Collection/Alamy Stock*)

locomotive driver but her mother's ill health meant that she had to stay at home to look after her following a series of strokes. Valentina's older sister was by then married and living away from home. Valentina's upbringing and education were typical of that generation who had come through the war as young children (she was 8 when the war ended). Education started when she moved with her mother from the countryside to the city of Yaroslavl in 1945. Life was hard and bereft of comforts – she recalls having no dolls or toys to play with. Education was valued: her mother was well read and raised Valentina to believe in the communist values that life would improve through individual contribution to the collective wellbeing and a belief in humanity's goodness. Having completed her statutory eight years of schooling by age 16, Valentina did not want to stay on and went to work in a tyre factory from which she transferred to the textile plant and enrolled in a technical institute for further education at night school. She graduated from the Textile Technical Institute as an engineer in 1960, by now fascinated with machines and engineering. Valentina's other great interest was the sky and flying, and through the Komsomol, at the Red Perekop factory, she was able to join the DOSAAF aero club and learn to parachute and skydive. She excelled at these and thoroughly enjoyed the excitement and feeling of falling through the air.

One of the factors learned in the early space flights by both Soviet and American teams was the importance of taking into account the twenty-four-hour cycle of life – the basic need evolved over millions of years for humans (in simple terms) to sleep at night and be active during the day. As well as this, there is also a monthly one involving the moon and its effect on tides and other natural phenomena including human functioning, a feature more popularly understood in the nations of the Soviet Union than in the West. The obvious impact of the changes this brings is with a woman's menstrual cycle. With the women trainees, and plans for one of them to go into space, this too had to be taken into account as well as all the factors that affected the performance of the men. This was just one of the matters for deciding when, and with whom, the next flight should take place.

Although it appears at first sight that there was a planned sequence to the Vostok series of space flights, this was only partially the case. Technical delays and changes of plan due to the interest (or disinterest) of the leadership altered proposals and priorities. There were also various factions involved and although Korolev was a driving force, his poor health had begun to remove him from the scene for periods. The idea that a Soviet citizen should be the first woman in space was given some priority when Titov returned from a trip to the US in May 1962 where he learned in conversation with astronaut John Glenn that the Americans planned to get a woman into space on a three-day flight by the end of 1962. One of Korolev's initial ideas for the women involved two in the same spacecraft – a mission that would follow a solo woman flight and form the main programme for 1963. By late 1962, this had been overshadowed in favour of another dual mission involving two simultaneous flights of women who would spend a longer period in space. Under pressure from the men, including Gagarin, this was changed again to one of the simultaneous flights involving a male cosmonaut. Valery Bykovsky was hastily brought forward for this role, delaying the mission by a few months.

The women's training intensified as the reality of flight neared, and, according to Siddiqi, there was an element of competition among them that was absent between the men trainees. The men knew that a mission awaited them at some point even if it was not the next one. The women knew by the end that opportunities for them would be limited and this concentrated their minds. Final examinations had taken place in November 1962, by which time Tatiana Kuznetsova had dropped out for health reasons – having performed poorly in the pressure chamber and on the centrifuge. The other four all passed with high grades and were formally inducted as air force second lieutenant pilots (their training had involved many hours of flying jet trainer aircraft) and cosmonauts. Of them, Valentina Ponamareva and Valentina Tereshkova showed the most all-around ability. Ponomareva was highly intelligent, independently minded and prone to speaking her mind – not recognizable assets to someone of the old school like Kamanin. Tereshkova on the other hand was a quieter and more compliant individual, whose working-class 'good breeding' was recognized by him as a particular quality. With Khrushchev's approval, it was eventually resolved that Tereshkova should be the first woman in space, with Irina Solovyeva her back-up. Tereshkova, Kamanin commented, was 'Gagarin in a skirt' (Siddiqi, p381). However, in a plan reminiscent of the story of the first flight of Gagarin, it is said that Tereshkova also fitted the bill for the first woman's flight as it would be highly automated, little would be expected of her, and she would make an ideal ambassador for Soviet women in the aftermath because of her political background. The respective skills of Ponomareva and Solovyeva would be reserved for the more complex dual flight that was to come: a spacewalk was planned that would particularly suit the physical strength of Solovyeva, with the skills of Ponomareva deployed in the role of commander.

Bykovsky launched into space on 14 June 1963, followed two days later by Tereshkova. Bykovsky's flight of five days set a new record for time spent in space, while Tereshkova's lasted three days. Both landed safely at the end of their missions and were greeted with all the usual pomp and ceremony that marked this latest Soviet achievement. The extent of the minor problems that beset both missions only emerged over forty years later. Bykovsky's flight, which had initial problems (unknown to him) at launch, was cut short as his orbits were lower than planned. Tereshkova suffered physical issues similar to Titov's but did her best to fulfil the tests and tasks required of her, including a short period of manual control. At the end of her mission, she ejected from the capsule only to see that she was headed for a large lake, and certain drowning was avoided when the wind carried her onto dry land where she was recovered safely. After the mission, despite the worldwide acclaim and prestige her accomplishment achieved, her efforts were derided by some within the space establishment and there was even an attempt to damage her character with a false allegation of drunkenness and poor behaviour. Kamanin defended her vigorously and her opponents were successfully challenged. To the world, Valentina Tereshkova was a hero and she proceeded to perform in this difficult and demanding role for many years and in many settings with dignity and charm. She grew into the task – at the outset in 1963 Kamanin remarks in his diary that there were issues with her rather impulsive and temperamental personality and that fame had rather gone to her head. He was convinced however that she was better suited to this ambassadorial role than either Ponomareva or Solovyeva might have been. She was neither the cool party apparatchik some would characterize her as, nor the typically unreliable member of her sex viewed through the prejudiced eyes of some of her male colleagues. Valentina was lauded in Soviet propaganda and projected around the world as the ideal Soviet woman – which meant playing up her femininity and dumbing down her military associations. She became the poster girl of the Soviet's peaceful intentions with their space programme. Valentina desperately wanted to go into space again and continued for several years to train and prepare for such a role, but was too valuable an asset for such a risk. The would-be train driver and textile worker had come a long way.

1963 Soviet booklet *Our Seagull* – Tereshkova's call sign when in space. (*Author's collection*)

Soon after the flight it emerged that Valentina was in a relationship with Andrian Nikolayev – the only unmarried veteran cosmonaut. This was seized upon by the Soviet establishment as a propaganda gift and the couple were encouraged to marry, although there is evidence from Kamanin's diaries that Nikolayev was less than convinced that he wanted to commit to a future with this immature and headstrong young woman. The window for a wedding that suited everyone was a short one and Tereshokova was becoming embarrassed at the constant questions about the matter as she toured. Whatever changed Nikolayev's mind, at the last minute in late October 1963, the couple announced their wedding, which took place on 3 November with First Secretary Khrushchev himself presiding at a much-publicized ceremony in Moscow. In keeping with traditional Russian patriarchy that survived in Soviet times, Nikolayev was rapidly promoted to colonel so that he would outrank his wife who was a major. A month or so later Valentina announced that engagements would have to be curtailed the following year as she was expecting a baby. The couple's child, Elena, was born in June 1964. This celebrity marriage between two young people under immense pressure was not to last, however: the couple stayed together formally until their divorce in the early 1980s. Valentina subsequently remarried. She continued as an air force officer and celebrity communist until the end of the Soviet era and later re-emerged as a politician in the Russian Federation Duma as a representative of President Putin's United Russia Party. In March 2020 Valentina Tereshkova was a central figure in constitutional changes that would allow Putin to continue his presidency into the indefinite future.

The projected multi-woman flight was cancelled and it was to be another nineteen years before the next Soviet woman, Svetlana Savitskaya, undertook the first of two space missions. Valentina Ponomareva, Irina Solovyeva and Zhanna Yerkina were never to emerge from the shadows.

Wedding of Valentina Tereshkova and Andrian Nikolayev, Moscow, November 1962. (*A. Mokletsov RIAN via Wikimedia Commons*)

Soviet Space Rituals and Customs

During the war Soviet soldiers at the front were fond of ritual, the breaching of which was said to bring bad luck, and given its near certainty for many anyway, death in battle. The cosmonauts also faced death with each mission and they too, from Gagarin's time onwards, adopted shared customs that were said to bring luck. According to Vostok pilot Leonov, the crew would spend the night before a flight in a particular house near the Baikonur base that had been used successively by Gagarin and Titov. The morning of a mission would start with the opening of a bottle of champagne, with the crew and others (including Korolev) each taking a sip, signing the label and then putting it aside to be finished on completion of the flight. Another common Russian custom before the start of a long journey is to sit down briefly for some self-composure before the person in charge suddenly stands up and announces, 'Okay, let's go.' Finally, on the bus journey to the launch site, the cosmonaut(s) would get out and urinate on the wheel – a custom followed by the women too, who are said to have preserved modesty by sprinkling a bottle of urine in the same place. Another ritual that developed with the advent of multi-manned crews was a pre-flight viewing of the 1970 Soviet movie *White Sun of the Desert*. This is said to have first been shown as an exemplar of technique for those charged with filming their flight and the views from space. It has since become a fixed tradition and to this day the space station

The Baikonur cosmonauts' bus to the launch site. (*Novosti* Man in Space, *1965*)

Soviet cosmonauts 1961–67. (Космонавтика – Маленькая энциклопедия, *1968*)

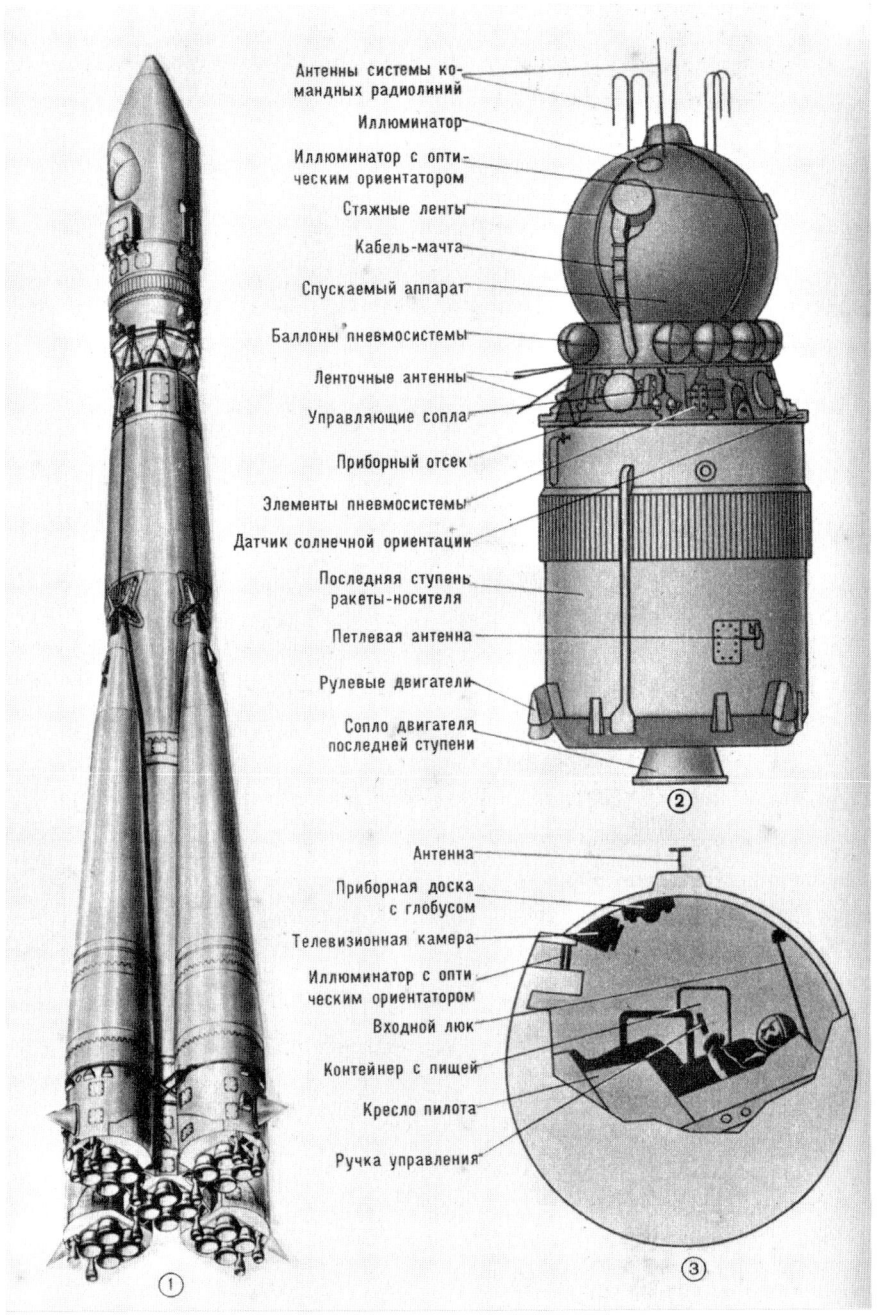

Vostok three-stage rocket. (Космонавтика – Маленькая энциклопедия, *1968*)

crews that fly from Baikonur, including those from Western countries like Britain and the US, are made to sit through this ageing, Russian-language, Soviet-era film.

* * *

The years between 1957 and 1963, which marked the high tide of Soviet success in space, were very much associated with Nikita Khrushchev. He understood the space programme's potential role in furthering the ideals of the Soviet Union, and a combination of circumstances ensured that adequate investment would result in success. These included the vision and brilliant leadership of Korolev, the initial absence of investment and interest by the Americans, and the links with rapid military missile development. That could not last – the Americans were fast catching up and background factors within the USSR were changing, including Khrushchev's removal from power in 1964. The next prize was a moon landing and the US had declared its intent to get there first. The Soviets gave up on this competition at an early stage but managed an array of other firsts, as we shall see in Chapter 7, before decline set into the Soviet space programme as the USSR's economy stagnated in the 1980s.

Valentina Tereshkova and Andrian Nikolayev in their Volga car, Star City, 1964. (*TASS/Valentin Cheredintsev Alamy Stock*)

Chapter 6

Selling the Dream

Yes, our Russia, our Soviet Russia is ahead of the game. Let some enemies beyond the seas look askance at us. They saw for themselves that they had to step aside (and they have done it lately). There are we ahead, the Soviet people. It cannot be in any other way. I pass great thanks and deepest gratitude to everyone who created the spaceship.

Excerpt from a letter to Yuri Gagarin from Uzbek Republic teacher L. V. Shulichenk, taken from *Ships Go into Space*, 1967

In the US there was always transparency about the space programme based on the historical workings of American democracy. While some military matters were secret for obvious security reasons, taxpayers and politicians took it for granted that they should be informed about how their money was being spent – especially on big-ticket items like space technology. While much of the technical details of the space programme might be found in scientific journals, a lot could be gleaned from popular mediums like children's comics. Not so in the USSR where secrecy dominated. Broad details of future plans might appear in the press, but any precise information about space flights and the personalities involved in them would only be released upon successful completion. Even then technical details that might be copied were jealously guarded. When the first painted impressions of the Vostok flights were released in the USSR in the early the 1960s, they were based purely on the artist's imagination. Even then, changes were ordered if they looked too much like the real thing. This resulted in a rather contradictory approach, given the recognized importance of showcasing success in space as being a direct consequence of the creation and ideals of the Soviet state: secrecy had to be aligned with celebration of achievement. One outcome involved the growth of a nascent consumer industry unlike anything seen before in the USSR. Showcasing also involved intense activity abroad – both in the capitalist countries as well as the friendly ones. The forms all this took, how it was managed and what impact was made, are the subject of this chapter.

ROCKET LAUNCHING

An artist's impression of a Soviet three-stage rocket at the launching site.

1. Sputnik
2. Ejector mechanism.
3. Electrical battery.
4. Helium pressure to feed fuel to engine.
5. Fuel tank for second stage.
6. Oxydiser tank.
7. Second stage engine.
8. Electrical battery.
9. Fuel tank for first stage.
10. Oxydiser tank.
11. Turbo-pumps.
12. First stage engine.

A quite imaginary drawing of the Soviet launch site from 1958. (*Novosti* Soviet Sputniks)

Cosmonaut Envoys of Socialism

Soon after Gherman Titov's ground-breaking space flight in August 1961, a crisis erupted on the border between East and West in Germany. In preceding years tensions had mounted, especially in Berlin, over the movement of people and goods across the border between the two very different worlds that divided the city. Economic instability in the East had increased the number of people moving to the West, undermining the credibility of the East German regime. Eventually Khrushchev agreed to the closure of the border, which was publicly announced as it took place on 13 August 1961. Issues concerning criminality surrounding border crossings were now replaced by ones of competing ideology and loyalty. In a managed attempt to contrast socialist achievement with Western decadence, Titov was brought to East Berlin on 1 September, just a few weeks after his flight. His enormous worldwide prestige was used to justify the border closure as a measure to retain the purity of achievement and progress of the Soviet Union and its friends. As an ambassador of Soviet values and progress, this was a role that was naturally expected of the cosmonauts and Titov would never have questioned its importance. In a speech in front of a very large and enthusiastic crowd, he praised the German Democratic Republic and the achievements of its people and was later reported in the press to have praised the new measures to 'protect the border' and that 'the plans of the West German and international imperialists, who want to interrupt the building of socialism, have been dealt a powerful blow' (quoted in Gumbert 2011). He seemed confident in this new role and talked powerfully about the peaceful intent of the socialist nations, as represented by his mission, and the sabre rattling and aggression found in the West. The purpose of his visit was clear: to convince East Germans that hope and progress lay not in looking to the values and consumerism of the West, but eastward to the Soviet Union (there was significance in naming the rockets that now carried the cosmonauts *Vostok* – east). Rallying support and raising morale in the East was now an important focus for the veteran cosmonauts. So too was the showing off of Soviet superiority and peaceful achievement through visits to the West – a hearts-and-minds battle as important as any other in the Cold War.

The youth, vigour, and general likeability of Gagarin, Titov, Tereshkova, and their colleagues, represented the human face of Soviet achievement, but this had all begun after the first Sputnik flights a few years earlier. World fairs and 'expositions' have taken place regularly since the late nineteenth century, the largest of which are used to showcase the general manufacturing ability, scientific achievement and culture of the participating nations. At the height of the Cold War the 1958 Brussels World Fair, the first such event after the Second World War, had a particular significance in the ideological battle between the Soviet Union and the US and was approached very seriously by both sides. Preparations in the USSR began in 1955 when Sputnik was still little more than an idea. One observer has noted that the Belgian organizers used Cold War tensions to heighten the impact of the event by feeding information about plans to each of the protagonists so that their respective efforts might be stepped up (Siegelbaum 2011). The Soviet plan was bold indeed, and its

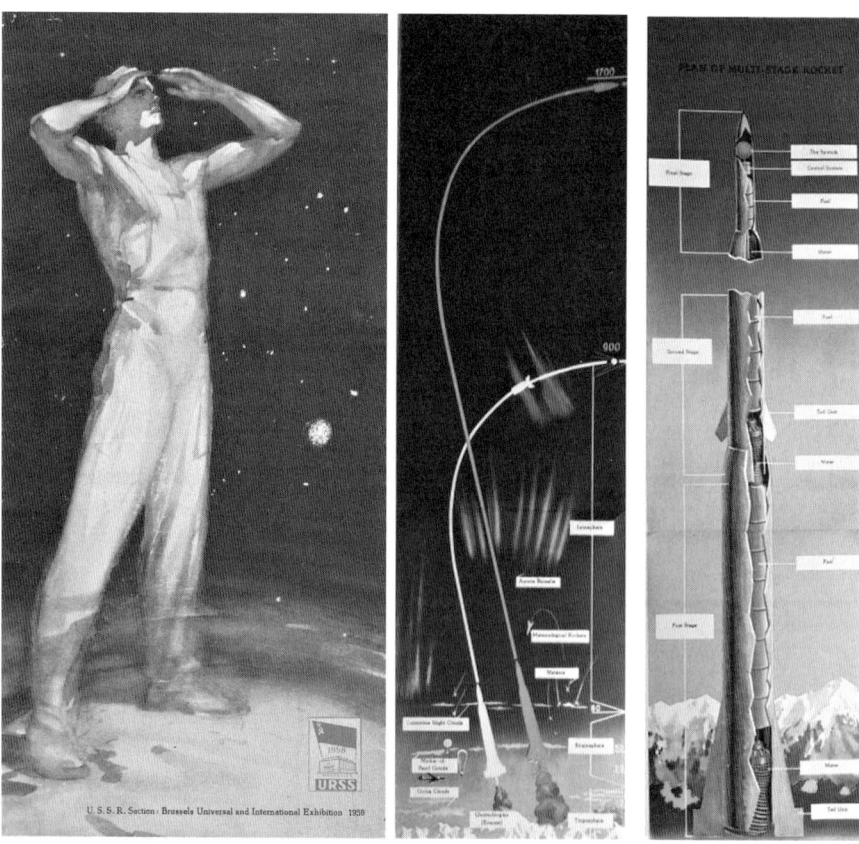

The cover and inside images of Sputnik from the Soviet brochure for the Brussels World Expo of 1958. (*Author's collection*)

original size would have dominated the exhibition but was scaled down to 269 square feet at the request of the organizers to match the floor-space dimensions of the US – which was located so that the two faced each other directly. The US's exhibition was designed to project the American way of life through its array of consumer items and icons of American success. The Soviet one, in which the public was met, on entering, by a large statue of Lenin, was also an attempt to project Soviet humanity through images of its people at work and play. However, its main impact and focal point was a life-size mock-up of Sputnik 1, surrounded by information about other technical achievements, including the nuclear-powered icebreaker, *Lenin*. One of the main inspirational architects of the USSR exhibition, Vasili Zakharchenko, personally designed a colourful brochure, reproduced in multiple languages for the Sputnik exhibit, which carefully avoided technical detail. The Soviet exhibition, due in no small part to the massive public interest in Sputnik, was a huge success, and one of the highlights of an experience which attracted over 40 million visitors.

A presence at international exhibitions, including at the beating heart of capitalism, New York in 1959 and London in 1961, continued to feature in the years to follow. These were always accompanied by high-

Brochure covers from trade exhibitions in New York (1959) and London (1961). (*Author's collection*)

quality publicity material. Indeed, Moscow-produced foreign-language promotional literature was churned out in large quantities throughout the period of the Soviet space programme. The international public face of the programme was the individuals who specialized in space matters from the Soviet Academy of Sciences. These were learned professors with a wide knowledge, but not directly involved in the programme. The literature they contributed to, along with the esteem of their names, especially in the early years that are the subject of this book, are notable by their generality and avoidance of detail. Novosti Press Agency material (some examples of which are listed in the bibliography) fits this description. A reader might be impressed by the message of technical prowess but would learn little of how this was achieved. One, a booklet from 1975 about the Baikonur Cosmodrome, is a late and rather bizarre example. Its twenty-six pages are extensively illustrated, but with only two images showing actual launches and any view of the site itself. The content is so bland that it struggles to fill the space provided. There is absolutely no indication of where in the Soviet Union Baikonur might be located, even though by this time it had long been known to the Americans. These booklets typically focused on pictures of dogs before 1961, and after that, pictures of cosmonauts (and their families) in various aspects of personal and working life, and public appearances. As such, they tell us more about the background to the times than they do about what happened in them.

As remarked upon previously, some of the leading figures, like Korolev and Glushko, did write for USSR-based publications, but anonymously, and this was not to change until later in the period, after Korolev's death. Books from the state publishing houses (who enjoyed a monopoly) for the home audience were cheap and plentiful (the Soviet Union was the biggest producer of books in the world), but also lacked technical detail during the period of success in the late 1950s and early 1960s. These books were generally printed in runs of 50,000 and were often compendiums of articles that had appeared in the Soviet press (some pamphlets, based on these, also appeared and would have enjoyed wider circulation). The general tendency was to fill them with idealized background details of the cosmonaut heroes. Examples of this genre are the lavish 761-page *Morning of the Space Age* from 1961 and *Cosmonauts Tell Their Stories* from 1964. By 1968, and the publication of *The Little Encyclopaedia of Cosmonautics*, more detail was permitted and its 528 pages are packed with facts about space exploration, including about American achievement. This book has information about Korolev – permissible after his death in 1966. If discussed at all, the competition from the USA might be derided, and the reaction of American leaders to Soviet successes used to underline Soviet superiority in the space race.

Ships Go into Space (1967) went beyond the usual triumphal journalism into a more sophisticated level of analysis, including a critique of the American space programme that takes up a whole chapter. This contrasted the skills and bravery of the American astronauts themselves with the profit motive that lay behind their organization. It details the involvement of ex-Nazis in the programme, the role of religion and the interest of the US government in space wars' programmes. In the early 1960s, the US was divided over calls to end gross racial inequalities in the southern states; moves to include a

Gagarin's triumphal entry to Moscow, 1961. (Утро Космической Эры, *1961*)

Titov's parents read about his space flight. (Утро Космической Эры, *1961*)

black pilot among the astronaut trainees were said to have been thwarted by reactionary opinion. A comment by Carl Rowan, an assistant secretary of state, about plans to provide equal opportunity in space, is quoted gleefully by the authors: 'It's all chatter because America can't even send a bus to Mississippi for black and white children to ride to school together.' (This theme was taken up by Black American songwriter Gil Scott-Heron in 1970 with his spoken word poem *Whitey on the Moon*, which compared the poverty of urban ghetto dwellers with the moon landing.) Elsewhere in the book, the accomplishment of Gagarin and Shepard, the first and second man in space, are compared and the conclusion is drawn that Soviet success was based on service to the state and people rather than to big business.

The rivalry was genuine but its public face was conducted at the level of superficial propaganda. Soviet cosmonauts and American astronauts had great personal respect for one another based on their shared experiences and the hardships of their training and preparation, paving the way for later joint programmes and cooperation. Their undoubted loyalty to their respective regimes did not stand in the way of their enjoyment of one another's company and the sharing of experiences. Soviet cosmonauts were generally careful not to give away anything very important – retribution would have been swift if they had, but as we saw in the last chapter concerning Titov's report about American plans to get a woman into space, US astronauts were not bound by the same level of secrecy. There were lapses in Soviet discipline. On a visit to Japan in 1966, Komarov revealed some unauthorized detail about immediate steps towards a moon landing and was severely reprimanded (Kamanin 1995–1997).

Some of the canine scouts were presented to the public, but after Gagarin's pioneering flight in 1961, Soviet emphasis at home and abroad focused on the human presence of the cosmonauts in an ambassadorial role. They visited every part of the USSR and most parts of the world in person and featured everywhere through images and stories in newspapers and popular magazines, becoming household names. Strict secrecy about who they were was maintained until missions were underway and assured of success. This was to the point of vague and erroneous information being fed to the Western press in the period leading up to Gagarin's flight – the normally well-informed British *Royal Air Force Flying Review* magazine for November 1960 printed the names and photographs of Soviet cosmonauts in training who, if they existed at all, were not in the cohort of trainees. One named individual, Vladimir Ilyushin, who was certainly real, was a famous test pilot and son of a celebrated aircraft designer. He was later the subject of rumours that he was launched into space prior to Gagarin in April 1961 but, as the tale goes, his flight went wrong and he landed in China where he was held by the authorities for a period. Stories about this appeared in the foreign press just before Gagarin's flight, including in the British Communist *Daily Worker*. They were later convincingly disproved – although they continue to feed conspiracy theories. It seems highly probable that they were initially spread by those in the Soviet establishment who favoured military space programmes over prestige projects and wanted to undermine the achievement of Gagarin and Korolev.

There was still, throughout the 1950s and most of the 1960s, a weariness about sharing even bland images of the apparatus involved. Soviet advances in aviation had always been demonstrated at the annual Aviation Day public air displays at Tushino outside Moscow. This at times involved imaginative enterprise to confuse watchful intelligence eyes from the West. In 1955, eighteen newly introduced 3-M Bison bombers were flown round twice in groups of different sizes, with

Young pioneers dressed up as cosmonauts, c. 1960. (*Author's collection*)

Cosmonaut Phillipchenko at the Oleg Koshevoi Pioneer Camp, Crimea, 1975. (*Author's collection*)

carefully obscured, but very long identification numbers, to make it look as if there were more than in actual existence. This suggested high levels of production and successfully made the Americans grossly overestimate the growth of the Soviet bomber fleet (Heppenheimer 2002). In July 1961, a large helicopter was flown into Tushino carrying what was said to be the final stage of a Vostok spaceship. This was mainly fictitious as what was shown bore only a vague resemblance to Vostok, not least because of its prominent, but purposeless, aerodynamic fin structure at the tail end (Siddiqi 2011). There was no need for such fiction with the cosmonauts – they were the real thing and what they said in public could be managed even if their behaviour in private was at times as flawed as that of most human beings.

Gagarin's flight quickly transformed him from unknown cosmonaut to sought-after celebrity – both roles closely managed by Kamanin. He was soon hurled into a hectic schedule of public appearances, press conferences and worldwide tours. In among all this, he found the time to continue active involvement in the space programme, at least for a while, personally answer fan mail and maintain a semblance of family life with Valya. A

'Vostok' at the Tushino Air Day, Moscow 1961. (*Unknown origin*)

world tour commenced at the end of April 1961 with visits to the friendly East European countries of Czechoslovakia and Bulgaria, and then to Finland. After returning to Moscow in June, Gagarin met several selected journalists – including Burchett and Purdy from the British *Daily Express* who went on to write an early book about his flight and life. Gagarin was adept at answering questions diplomatically and with charm, being careful not to give away any more information than was permissible and studiously avoiding technical details about problem areas or matters such as the real nature of his landing. In July he was back on tour – to London for an ambiguous official reception that seemed to have been arranged at short notice to keep up with public opinion, rather than by prior arrangement. This included the meal with the Queen and a meeting with the Conservative Prime Minister Harold McMillan, while he travelled around London in a Rolls-Royce with a personalized numberplate loaned to him for the occasion. Gagarin was met everywhere by rapturous crowds. In August he undertook a tour of Canada, which was cut short so that he could return home for the celebrations following Titov's successful mission. At the end of the same month he was off on a symbolically important tour of Cuba – the Soviet Union's protégé on the USA's doorstep.

After the Crimean incident, Gagarin's tour recommenced with visits across the world, taking in India, Egypt, Greece, Cyprus and Japan, part

Castro and Gagarin, Cuba, June 1961. (*Unknown Cuban photographer via Wikimedia Commons*)

of which he was accompanied by wife Valya. On his return, Gagarin was given a role in the selection and training of the women cosmonauts, and during the remaining years of Korolev's life was given increasing responsibility in the cosmonaut training centre at Star City. In among all this, Gagarin found time to personally answer the letters he received from people at home and abroad. There is evidence that he put thought into these – his cult status was such that fan mail included requests for advice on wider issues, and Gagarin did his best to extol the ideals of the new communist man and prove that duty and obligation were supreme values that he embodied as a pioneering cosmonaut. Soviet popular books published these letters – for obvious reasons focusing on ones that promoted Soviet values and achievement. Fifteen-year-old Canadian Irving Lazar of Montreal asked him several philosophical questions including if it could ever be right to tell lies, and what was the meaning of success. Gagarin's lengthy and carefully constructed answer states his belief that it was wrong to lie for personal interest (leaving other matters open) and expressing the hope that this young person would never have to do so in the future. Gagarin tells him that success has to be worthy of the struggle to achieve it, and for that you need comrades around you who share your beliefs (*Ships Go into Space*, 1967).

Many were from Soviet citizens, and one is worth quoting in full:

Dear Yuri Alexeevich!

You get many letters with congratulations for your flight around the globe, but I am sure you have no other written by a person without hands. I am just such a person. Both my hands are missing and I have no feet either. I have got 14 wounds on my body and a strong desire to live. I got disabled defending our Motherland from fascism. The last battle I took part on my feet, with a rifle in my hands, was on the 17th of December, 1942 on fields of the Volgograd region. Now I am a disabled person, but I am writing this letter myself, without anyone's help and I am writing to you – the person, who opened a new era, an era of flights into space.

Thank you, Yuri Alexevich! You and your sky brothers glorified our Motherland forever. It cannot be forgotten – it's our story and glory!

Thanks to your parents, who raised such heroes. Thanks to your wives, who supported you on so great a mission.

I am 58, my wife is 55. We have got two children: Margarita, 20 years old, a checker inspector in the plant, and a son of 33 years old (he is also Yuri), an engineer in the plant who is married, and we have a Grandson. Our children are happy and well.

After the war I was a deputy of the district Soviet in Perm, and worked as a senior bookkeeper in the social services office. By my labor I provided benefit to the Motherland. In 1957 I was conferred the rank of personal pensioner of the Russian SFSR. I have lived a full and interesting life. I have got a good flat with gas, hot water, heating and a telephone. We have got a radio, a TV-set, a refrigerator, piano, books, newspapers and magazines. My son possesses a 'Moskvitch' car. I and my family have got it all, because we live in the USSR. My dream is to live to see communism.

Anton Pustozerov.

(From уходят в космос корабли, *Ships Go into Space*, 1967, p245–246)

Gagarin and several of his colleagues penned a collective reply which began:

> Thank you, Anton Ivanovich, and your friends, comrades in arms and people of your generation, who by their lives and often deaths, prepared the deed of our generation in beginning the conquest of space. (Ibid.)

Gagarin's role in exporting the message about the Soviet space programme was continued with similar tours by the other Voshkod flight cosmonauts in the early to mid-1960s, and after this by the later Voskhod and other programme crews. They too were deluged with correspondence from well-wishers. Leonov (see Chapter 7) later recounted a letter he received from a professional burglar who compared the dangerous nature of space travel with its risk of death, to the risk of imprisonment in his own line of work, resolving to go straight in honour of the cosmonauts (Scott & Leonov 2005).

Gagarin welcomed in Warsaw, Poland, 1961. (*Unknown Polish photographer via Wikimedia Commons*)

Tereshkova was especially sought after abroad, there being a fascination with her as a woman cosmonaut, and when she toured with colleague Bykovsky, he was very much in her shadow despite the important achievement of his own flight. This interest in her as a woman was exploited to the full. Within weeks of her flight, she was hosting the International Women's Congress in Moscow, a gathering of over 2,000 women from 119 countries. Between 1963 and 1970 Tereshkova visited 42 countries, her continuous work as the outstanding representative of Soviet womanhood hardly ceasing for the pregnancy and birth of her baby. She visited Great Britain in February 1964 and met the Queen (who was also pregnant), just as Gagarin had done. In 1968, Tereshokova reluctantly put her dream of training for a further space mission on hold to become chair of the Committee of Soviet Women. The following year the female cosmonaut team was abandoned, which dashed whatever remaining hope she may have had, but she stoically continued with the official tasks given to her. Throughout the Soviet era, Tereshkova remained an active air force officer, and when the idea of women cosmonauts was revived in the late 1970s, she underwent medical tests to resume her role – this was denied her but she did remain involved as an instructor.

Valentina Tereshkova and daughter Elena, 1964 (*Everett Collection/Alamy Stock*)

Valentina Tereshkova and American Communists and black activists, Angela Davis and Kendra Alexander, Moscow, 1972. (*D. Chernov RIAN via Wikimedia Commons*)

Promoting Atheism

In Tsarist Russia the Russian Orthodox Church gained notoriety for its role in propping up the absolutist monarchy and their wealthy, privileged supporters, and convincing the country's poor (who were the vast majority of citizens) that this was their place in the natural order created by God. After the October Revolution, one of the aims of the Bolsheviks was to eradicate religion in the belief that this would free the people (especially the peasantry) from superstition and help them improve their lot. This was pursued vigorously by Stalin in the 1930s and then set aside during the war, before being resurrected in the 1950s and 1960s. At times the Orthodox Church was violently repressed: churches were demolished, and priests terrorized and imprisoned. None of this was especially effective in changing the attitudes of older people, but younger ones grew up with little notion of the concept of religious belief. This was the case with those of the cosmonaut generation, who were brought up to regard atheism as natural and religious belief as something belonging to the country's past. Some six to ten million people felt empowered and obliged to spread an atheist message – they were typically found among Party members and activists in schools, in the media, in factories and offices (Pankhurst 1988).

On the ideological front, the Bolsheviks put considerable thought into how to convince people that scientific progress and discovery had rendered religion obsolete. Knowledge of the planets and the world beyond Earth played a significant role as they contradicted some of the teachings of the Church. From the 1930s, planetariums were built in the major cities to popularize learning about the planets and natural order according to science – by 1973 there were more than seventy, many built symbolically on ground formally occupied by churches. Typical was the Kharkov Planetarium in Ukraine, which was opened in 1957 thanks to the work in the city's university of astronomer Nikolai Barabashov. It was he who calculated, based on observation, that the moon's surface was solid and could take landings of spacecraft. An acknowledgement of his work was later given by American scientists after the successful

The Kharkov Planetarium bus and mobile telescope, 1957. (*Kharkov Planetarium*)

Outdoor education on a collective farm, 1959. (*Kharkov Planetarium*)

Apollo 15 moon landing. Kharkov Planetarium later gained a powerful Zeiss telescope and remains an important visitor attraction and centre of study. From the outset, it employed a new bus fitted to provide a mobile planetarium so that a message of natural scientific knowledge could be carried to ordinary citizens, including peasants on collective farms – its underlying purpose was the promotion of atheism. The planetarium itself had a large meeting hall where ideological, as well as scientific lectures, were offered to the public.

The popularity of the first cosmonauts lent a new impetus to the promotion of atheism, and they were charged with spreading this message to Soviet citizens. They, after all, had broken through some of the mysteries that surrounded religion and could say with certainty that everything they had seen and experienced could be explained by science. This was the theme of an *Izvestia* newspaper editorial in May 1961, shortly after Gagarin's flight, which attacked religious belief by pointing out that he had seen 'neither the Almighty, nor Archangel Gabriel, nor the angels of heaven' (quoted in Smolkin-Rothrock 2011, p165). Titov was widely reported in the West when he commented at the Seattle World's Fair in 1962 that he had he seen no evidence of a deity in space and that he did not believe in God, but 'in man, his strength,

Lecture in the Star Hall of Kharkov Planetarium, 1957. (*Kharkov Planetarium*)

his possibilities and his reason' (Ibid., p166). However, as time went on, cosmonauts found that debating such notions with religious believers, and fielding their questions, was far from straightforward. Religious belief is a complex matter, and the crude Marxist notion that it would disappear in the wake of human progress proved problematic in practice when the focus was solely on scientific evidence. The cosmonauts realized that having been brought up as communists they knew little about religion and this had some unexpected outcomes. Titov suggested that cosmonauts study the Bible to better understand the basis of Christian belief; Gagarin criticized the demolition (in the 1930s) of the important Church of Christ the Saviour in Moscow and received applause at the Central Committee of the Communist Party when he did so. Eventually, the Soviet state's strategy of combating religion changed, and by the late 1960s, the case was being made through philosophical, rather than scientific, argument and the cosmonauts quietly slipped out of their role as advocates for atheism.

Pin Badges, Stamps and Other Souvenirs: The Spread of Cosmic Consumerism in the USSR

The Bolshevik Revolution was built around the needs of working people and peasants. Although many of its leaders came from bourgeois backgrounds, the Communist Party in the early years after 1917 eschewed the preoccupations of the wealthy, such as the private accumulation of personal possessions that did not meet immediate needs. The personal consumption of goods produced by the workers' state was for the essentials of life – food, clothing and household items. Surplus wages might be spent on subsidized sporting activities conducted collectively, books, or even holidays, but in general, items of aesthetic value only, such as works of art, were held in museums and galleries for public enjoyment. There was no place for the widespread Western practice of 'collecting'. This began to change as early as the 1920s, in limited form, with the circulation of postage stamps and small metal pin badges – echoing back to Tsarist times. Both were cheap and occupied little space in overcrowded homes. Pin badges, commemorating important anniversaries, events and places, also mirrored the state's many awards for labour competition or war service that would be worn by the holder.

The prestige of the Soviet space programme was important to consolidate the belief of ordinary citizens in the values and achievements of their state. Although thousands were involved directly or indirectly in its design, manufacture and manifestation, most Soviet citizens could only view the space programme from a distance – it could never tap the sense of mass involvement and unity of purpose felt by Soviet citizens in the Great Patriotic War. Such though, was the weight of cultural propaganda around the space successes from Sputnik onwards, especially after Gagarin's flight, that Soviet citizens wanted to identify with the technology and the heroes who were involved. Stamps were a utility in mass use so their change to commodity status in the form of collecting was simple, and took nothing away from the productive effort. It had been authorized (and therefore controlled) since the 1920s with the regular output of artistically designed stamps propagandizing important events and achievements, and was even encouraged on a small scale with the limited publication of catalogues. In the 1950s, stamp designs changed from the short slogans and socialist realist images of the Stalin

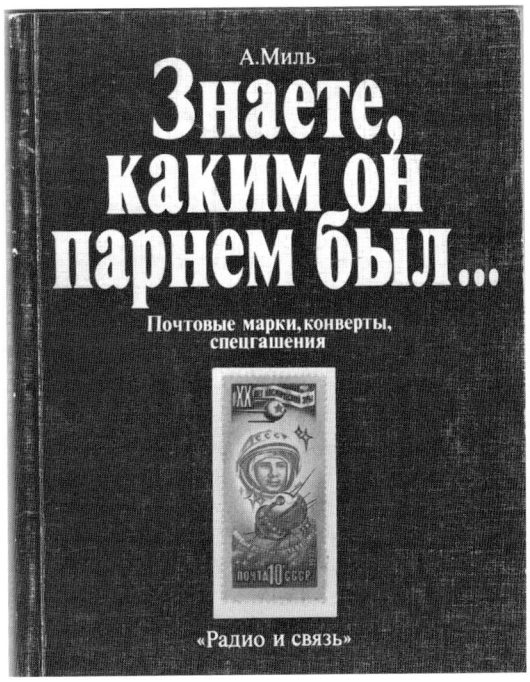

Soviet book about space-themed stamp collecting from 1984. (*Знаете, Каким он Парнем Был*)

era to more detailed reflections of Soviet policy, making them arguably less attractive. Postage stamps celebrating the Sputnik successes followed these styles and continued (including special envelopes) after Gagarin's flight, nonetheless giving a boost to philately in the USSR. The designs became more interesting from about 1965, and by 1975 more than 100 space-themed stamps had been issued (Lewis 2011). This relaxation of regulation concerning hobbies accompanied some concerning contacts with the outside world (through organized activities rather than freedom of movement) – both aspects of the liberalization of the Khrushchev years. Collecting as a private pastime received official encouragement from the early 1960s – aiming to encourage interest and a feeling of involvement by ordinary citizens in the space programme.

The wearing of officially produced pin badges to commemorate events and activities never ceased and included a number celebrating the Sputnik programme after 1957. However, after Gagarin's flight, the whole idea snowballed and pins connected with the personalities and achievements of the space programme became so numerous that their numbers and designs are said to be impossible to document. Many state enterprises,

'Znatchki' pin badge examples. (*Author's collection*)

who used metal processing in production, turned over surplus capacity to pressing pins with space images and slogans – some simple and some more complex, and often produced in limited quantity. (This is of interest as such production lay outside of the centralized planning that governed production in general terms.) While pins were worn by anyone wanting to identify with the cosmonaut heroes and space successes, they also quickly became sought after by a new generation of collectors. Costing about the same as a loaf of bread, they were bought up by amateur collectors. Unlike Western collecting where rarity is sought after, Soviet collectors favoured the most popular and widely available items – mirroring perhaps pop and fashion culture in the West. In the 1970s books appeared that detailed designs and boosted collecting as a pastime in the USSR.

Other memorabilia reflecting the space programme appeared in varying quantity. Some were relatively expensive and designed to adorn the desks of state officials. 'Epoch' desk lamps, made in the early 1970s simulating a colourful rocket launch when switched on, were made in the Yuzhmash plant in Dnipropetrovsk, Ukraine, that produced space equipment. Beyond the reach of ordinary citizens, these might be given as awards or gifts. Plastic (sometimes metal) statuettes and pen sets

Soviet space mementoes, clockwise from top left: cigarette case, Vostok desk ornament, spoon, manicure set. (*Author's collection & Valentina Kudinova*)

in socialist realist style also had the same limited availability but were advertised abroad – in 1961 the English magazine *Soviet Union* described them as 'by far the most precious, for they have been made in the land of the first Sputnik and the first space pilot'. Ordinary citizens were not left out – cheaper household items with a space theme ranged from cutlery to manicure sets and cigarette cases. All are now of interest to Western collectors and fetch keen prices on internet auction websites.

Space in Popular Culture

Movies, as a reflection of Soviet society, were discussed in Chapter 1. Heroes had always featured large, and this was certainly demonstrated in the huge volume of films made about the Great Patriotic War, from the outset of the war itself through to the end of the Soviet era, and beyond. Space received a different treatment, probably because of the ongoing secrecy surrounding the programme that precluded its accurate representation on the screen. However, the Soviet dream of space travel received treatment from the earliest days of Soviet cinema in the 1920s, reflecting the nation's fascination with science fiction. The IMDb online movie database lists twenty-eight sci-fi movies, and four documentaries, made between the launch of Sputnik in 1957 and the end of the Soviet era in 1991. There were no cinematic representations of the space successes but the drama documentaries covered the life of Korolev (*Taming the Fire*, 1972), Gagarin's childhood (*So the Legend Began*, 1976) and two concerning Tsiolkovsky (*Man from Planet Earth*, 1959, and *Take Off*, 1979). Since the end of the Soviet era, in contrast, there have been numerous Russian movie dramas with themes based on the space race. It seems again that Soviet society was hamstrung by secrecy that prevented using this medium to celebrate success.

Of the sci-fi movies, the most internationally famous is director Andrei Tarkovsky's *Solaris* of 1972. Unlike Western movies of this era and genre, which focused on imaginary technical innovations, whose characters lacked much depth, *Solaris* delves into human psychology and reflects some of the concerns of Soviet scientists as they pushed boundaries within the real space programme. Its plot concerns a three-man-crewed space station orbiting a distant planet, whose inhabitants have all become affected by a deep emotional crisis. A psychologist is dispatched from

Earth to investigate but falls prey to the same mysterious illness and a complex and elaborate story ensues which examines the inner lives of the characters. The film won acclaim worldwide and is still regarded as one of the best sci-fi movies ever made. After a cautious and narrow release in the USSR, the film went on to run on a limited basis for fifteen years – of artistic appeal to a small rather than mass audience (the aspiration of most Soviet movies). Tarkovsky's film achieved lasting cult status (Lawton 1992).

The space race was also imaginatively celebrated in other art forms – the visual arts followed the movies in taking a sci-fi view because accurate representations of Soviet space technology were forbidden. The imaginative and colourful paintings of artists like Avatov, Pobodinsky, Sokolov and the cosmonaut Leonov (see Chapter 7) adorned books, popular magazines and postcards. Soviet space achievement was also represented in the popular song 'Fourteen Minutes to Launch', written by Oscar Feltsman and Vladimir Voinovich in 1960 by government order (in time for the first Vostok launch in 1961), and became the anthem of the cosmonauts.

A young soldier's personalized New Year 1964 greeting card. (*Author's collection*)

> Packed into the tablets
> Are the space maps
> And the Navigator checks
> The route for one last time.
> Let us, guys,
> Smoke before launching:
> We still have fourteen minutes
> Left until the launch.

(Chorus)
I believe, friends,
Caravans of rockets
Will head us forward –
From a star to a star.
On the dusty paths
Of the distant planets
Our footprints'll be left as our marks.

And after many years
We'll remember with the friends,
How on the stellar roadways
We were the first to venture,
How we were first to manage
To reach the cherished goal
And from the greater distance
Look at Mother Earth.

Long are we awaited
By the distant planets,
Cold worlds,
Silent fields.
But not one of the planets
Is waiting for us as this
Precious planet
Named the Earth.

Its most famous rendition was by the popular singer Vladimir Troshin but was sung in space from Vostok 3 by Nikolayev and Popovich. It can be easily found and heard online.

* * *

The 1960s and early 1970s mark the heyday of space propaganda in the Soviet Union. This was manifested in various ways to maximize the impact of Soviet achievement in the space programme both at home and abroad. The aspect that achieved the most success across the world was the ambassador role performed by the cosmonauts themselves. The flights and their subsequent publicity were accompanied by literature and by a growing consumer industry of assorted items and memorabilia associated with the space programme. Such a development marked a small but significant new beginning of non-essential consumerism for the Soviet Union that would continue, with results that undermined the goal of communism, into the 1980s. This will be discussed in the final chapter.

Chapter 7

The Race to the Moon: Failure and Retreat

Uncle Kolya was in Moscow on holiday. He decided to call on the Smirnovs but everyone was out except little Nina.
'Where's your daddy?' he asked.
'He's building a satellite station in outer space. He'll be home in two hours' time.'
'What about your brother Yuri?'
'He's on a moon probe. He'll be home later this evening.'
'And your mummy?'
'Oh, she'll be ages. She's queuing up for a new pair of shoes.'

1970s popular Russian joke

A few days after Gagarin returned to Earth, President Kennedy sent a memo to his vice-president Lyndon Johnston in which he asked:

1. Do we have a chance of beating the Soviets by putting a laboratory in space, or by a trip round the moon, or by a rocket to land on the moon, or by a rocket to go to the moon and back with a man? Is there any other space program which promises dramatic results in which we could win?
2. How much additional would it cost?
3. Are we working 24 hours a day on existing programs? If not, why not? If not, will you make recommendations to me as to how work can be speeded up?
4. In building large boosters should we put our emphasis on nuclear, chemical or liquid fuel, or a combination of these three?
5. Are we making maximum effort? Are we achieving necessary results?

(Quoted in Smolders 1973, p38)

The Race to the Moon: Failure and Retreat

Within a few weeks, on 25 May 1961, Kennedy announced to the world that the Americans would be on the moon by the end of the decade. While this effectively acknowledged the likelihood of continued Soviet superiority in some areas (such as the possibility of space stations and laboratories), the advice was that eventual American success could be achieved by sheer spending power in a prestigious attempt to be the first to place a human being on the surface of the moon. Interplanetary travel, and not just limited to moon exploration, had always been an objective for Korolev's plans – not just in his imagination either. The Soviets were keen to rise to a challenge they felt they could win, and the much-discussed and reported race to the moon was on. The Soviet successes, as we have seen, were built upon a combination of factors, including the political support and enthusiasm of Nikita Khrushchev. He realized, more than other Soviet leaders of his generation, that space success helped build public belief in the Soviet state, prestige abroad and, at least initially, the development of military ICBM capacity. However, the Soviet supremacy that accompanied the 'firsts' of the 1957–1963 period was coming to an end, and momentum was to be lost by the end of the decade. How that happened is the subject of this chapter.

NASA press conference 'welcoming' Gagarin's flight. (*NASA via Wikimedia Commons*)

The Lunar Programme Gets Underway

Soviet moon exploration began as early as 1958 – and unmanned robotic spacecraft were sent into space on moon missions between then and 1976. The first success was in September 1959 when a missile successfully crashed onto the moon's surface. A month later a satellite orbited the moon and took photographs of the far side. A series of orbiting missions took place in 1966, but major success preceded them in January of that year when E6-No. 13 (known publicly as Lunar 9) landed intact on 3 February 1966, followed by similar success with the E6M-No. 205 (Lunar 13) in December 1966. The next successes were in 1970 with missions that returned to Earth with samples and then deployed a vehicle for exploration (Lunokhod 1) – after the success of the US Apollo 15 flight which took men to the moon and back. There were further similar missions before the programme was abandoned in 1976. In all, fifteen out of forty-four unmanned missions resulted in success – the failures being hidden from public view. Korolev began plans for a manned moon mission in 1959, and although this was given political imperative after Kennedy's announcement, little actual progress was made. By the mid-1960s the Soviets were thought to be some three years behind the Americans in the development of the necessary technology.

Soviet supremacy until the mid-1960s rested on the R-7 rocket which was powerful enough to support the Vostok and Voshkod spaceship launches. However, a spaceship capable of moon landing and re-launch required a much more powerful launching vehicle, and work on the N-1 system which would deliver this began in the early 1960s. Parallel with launch-vehicle development was the start of the Voskhod (meaning dawn or ascent) programme of space flights that would extend knowledge already gained from the six solo Vostok missions, and orbit further away from Earth. The Vostok flights had taught much about the effects of space travel on the human bodies of the very specially selected air force pilots involved. The training programme was now opened up to include those with other necessary scientific and technical skills, and the age range widened. Although this did not include the many thousands of ordinary Soviet citizens who enthusiastically volunteered themselves for space travel in the wake of Gagarin's and his colleagues' pioneering efforts, those involved came from a wider spectrum but shared loyalty to

The Race to the Moon: Failure and Retreat 135

the ideals of the Soviet state. The Voskhod 1 mission in October 1964, the first with more than one crewmember, involved Vladimir Komarov, a career air force pilot and one of the initial space trainees, as commander. Also aboard were a physician, Dr Boris Yegorov, and an engineering researcher, Konstantin Feoktistov. The capsules used were adapted from the Vostok design, with some safety compromises to enlarge space and omit the need for spacesuits during the flight. Technical advances were included to support their heavier weight and permit capsule parachutes

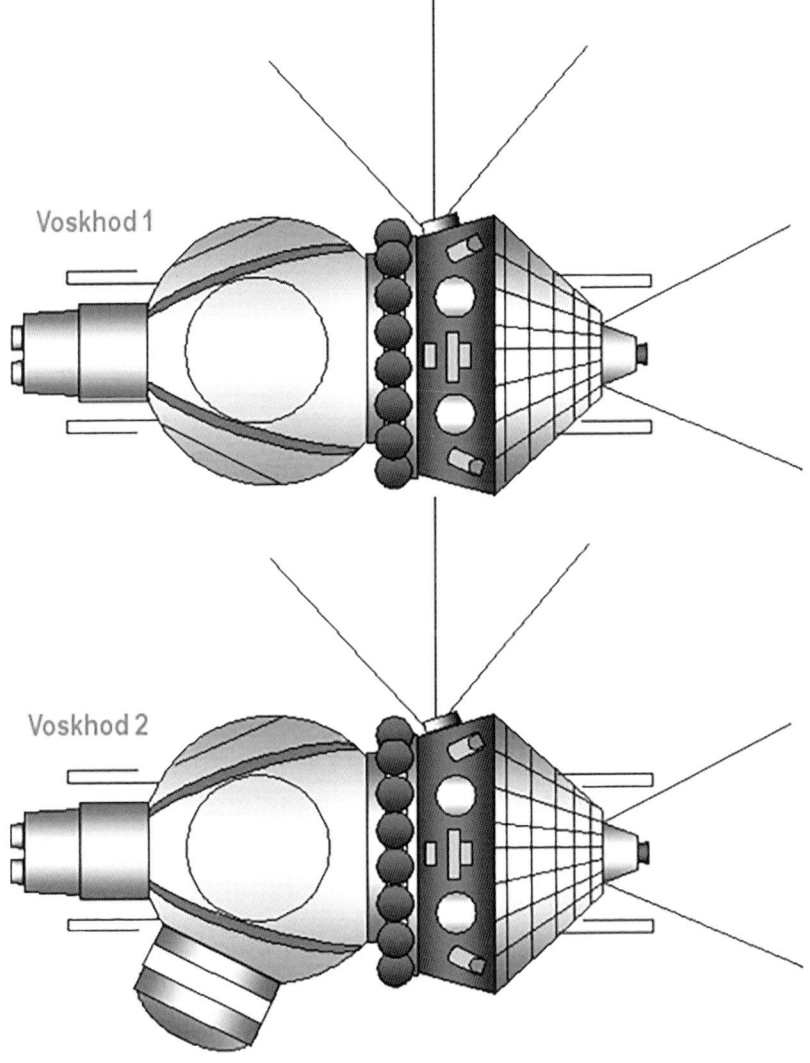

Voskhod 1 and Voskhod 2 capsules compared. (*Reubenbarton via Wikimedia Commons*)

landing with their crews still on board. The short-duration Voskhod 1 mission was a success and beat the American Gemini programme with the first multi-manned mission.

Voskhod 2 in March 1965, crewed by Alexei Leonov and Pavel Belyayev, achieved another spectacular result – a spacewalk by Leonov. Again, this beat (as intended) widely publicized American plans to do the same with astronaut Ed White in May that year. Although the world quickly learned of Leonov's ten-minute walk outside the Voskhod capsule as it orbited the Earth at 18,000 miles an hour, the actual problems of the flight, which almost ended several times in disaster, were kept secret at the time. At the end of his walk outside the capsule, Leonov found that he could not access the airlock chamber feet first as planned, and had to indelicately manoeuvre in with his head as his spacesuit had become loosened from his body in the weightless atmosphere. Unscheduled

Leonov in training for his first spacewalk. (*Novosti* Man in Space, *1965*)

(and untelevized) actions that could have been catastrophic. After the spacewalk, Leonov, an artist, relaxed by sketching what he could see with crayons and paper he had taken into space. He had hardly got underway when the devices for ejecting the airlock chamber sent the capsule into a repetitive roll at 17 degrees per second which the crew had to endure for a further twenty-two hours. While this was uncomfortable, but not dangerous, it was not the end of their woes. The oxygen pressure system malfunctioned and gradually increased to a dangerous degree, but later righted itself before re-entry was due to take place. Throughout this phase of the flight, and unbeknown to Leonov and Belyayev, the live commentary on Moscow Radio was replaced with the sombre music (Mozart's *Requiem*) that featured after the announcement of a Soviet leader's death. Clearly, a safe return was not expected. Re-entry to the Earth's atmosphere was also potentially disastrous when the landing module failed to properly separate from the orbital module, but again the situation was resolved just in time, and the capsule eventually landed far from its intended site, deep in the inaccessible Siberian taiga. It was located fairly quickly by a civilian helicopter that hovered above and offered to rescue the two heroes with a rope ladder thrown out of the side door – something that only happens successfully in fiction. They were eventually retrieved the next day after a massive rescue effort and flown back to Baikonur. It was to be many years before the near disasters of the mission were mentioned: as far as the world was concerned, this was another stunning Soviet 'first'.

Setbacks on Earth and Disasters in Space

Nikita Khrushchev's leadership came to an abrupt end in October 1964 when various elements in the Soviet leadership, who disliked his style and policies, successfully combined to replace him with Leonid Brezhnev, who would lead the Soviet Union for the next eighteen years until he died in 1982. Brezhnev, like his predecessor, quickly learned that the space programme, as a messenger of the Soviet Union's aim of peace and progress, as well as superiority over capitalism, was worthy of continued political and financial support. However, as the economy faltered under the burden of continued Cold War military spending (to which Brezhnev was committed) and the state failed to meet the basic needs (never mind the aspirational consumer demands) of Soviet citizens, aspirations were inevitably curbed and so too spending in the years to come. By 1965 there was already a feeling among the cosmonaut group that their significance was beginning to recede and in October Gagarin and five of his well-known colleagues wrote to Brezhnev asking for more investment on the basis that another 'first' be achieved for the 50th anniversary of the October Revolution in October 1967 (all referred to in Kamanin's diaries). The letter probably never reached its intended recipient and certainly never received a reply.

While this trajectory was inevitable, given the power of the American dollar to enable the US to catch up and eventually overshadow their competitors, the USSR's programme received a body blow from an all too human cause. Sergei Korolev had been suffering ill health for many years, which would not have been helped by the lasting effects from his time in the Kolyma gulag. He suffered a heart attack in 1960 and while recovering was advised by attending doctors that he had a chronic kidney disorder that would eventually take his life unless he reduced his work commitments. Korolev chose to ignore this, knowing that if the Americans started to overtake the Soviets, his plans would lose the political support that was the programme's lifeblood. Over the next five years, his various health issues escalated, compounded by fatigue from overwork. At the beginning of 1966 he was admitted to hospital for surgery and died nine days later. The actual circumstances of his death are disputed. The official version at the time was that he had an inoperable cancerous tumour, but other stories have emerged suggesting his death was the result of unintentional but botched surgery – the general conclusion drawn by his principal Russian biographer, Yaroslav Golovanov. It was at this point that the hitherto secret identity of the 'Chief Designer' was revealed in detail to the Soviet public and the world, and Sergei Korolev received a state burial in the Kremlin Wall alongside the USSR's most famous sons and daughters. Korolev was replaced by Vasily Mishin (1917–2001), one of his chief deputies and a long-time member of the design team.

The Voskhod programme was initially projected by Korolev and his associates to extend knowledge of the effects of spaceflight on the human body. However, due to the political imperatives of the space race, the study was reduced to the two 'firsts' of Voskhods 1 and 2. The remaining four missions were cancelled as their expected outcomes were all achieved within the rapidly advancing American Gemini programme. It was hoped that some momentum could be regained with the Soyuz (Union) programme that had been designed under Korolev's leadership. Soyuz spacecraft were a significant advance on the two earlier designs and consisted of three interlocking parts: a spherical orbital module, a smaller re-entry module and a cylindrical service module with solar panels for almost limitless power production. The ability of the modules to lock with one another in space was intended to allow a series of Soyuz craft to achieve a moon flight using smaller launch vehicles that could be made available by developing existing technology from the R-7 rocket. This might have provided a technically more viable system than a single launch using the large N-1, or other vehicles the size of the American Saturn rocket series, developed from the late 1950s with a moon landing in mind. The N-1, which was the most powerful rocket ever built, was still under development but was beset by problems, not least of which was Korolev's death. Rivalry existed between the two different systems, although both shared use of the Soyuz modules: the Soyuz multi-launch moon project would have involved UR-500 launch rockets designed by Vladimir Chelomey. Approval of the two systems was given in 1965 with an intended landing on the moon, by the most successful, in time for the 50th-anniversary celebrations in October 1967. The UR-500 also had a military application which justified these parallel developments. The N-1 system, now several years behind the American Saturn, lurched on and was not ready for launch trials until 1969. A hasty attempt to catch up with the imminent Apollo mission took place in early July and ended in a disastrous explosion at Baikonur that caused so much destruction that progress was set back for two years. The next two unsuccessful launches took place in 1971. A fifth was planned for 1974 when the moon programme was officially cancelled, long after the outstanding success of the Saturn-launched Apollo 11 mission.

The team behind the Soyuz system were under obvious pressure to deliver. Gagarin and Komarov, who had been assigned to the project, were critical of some aspects of the design, including those concerning safety, and felt things were being rushed. Engineers reported over 200 faults but delays were overruled by Party leaders who, after two years of Soviet human absence in space, wanted a series of 'firsts' leading up to massive celebrations for the 100th anniversary of Lenin's birth, in 1970. They were conscious that the American Gemini programme's numerous flights were rapidly overtaking the USSR. The first Soyuz flight was planned for April 1967 and would be commanded by Komarov, which would meet in space with a separate flight. Although all cosmonauts (and their American colleagues) approached each mission with some fatalistic trepidation, there is post-Soviet-era speculation (recounted in Doran and Bizoni 1998) that Komarov was sure that he was being sent to certain death and was angry about it. He is said to have insisted that he should command it and not Gagarin, who had been put forward for the mission, whom he regarded as too important to sacrifice so easily. However, Chertok disputes this account and insists that Komarov felt his close friend should not be denied the opportunity for further space flight. Kamanin's normally frank diary makes no mention of any doubts on Komarov's part and describes him exuding confidence about a mission that would make him the first person to enter space for a second time. He did not even tell his wife he was going away and she met the news after launch in a relaxed fashion. The plan involved meeting a second flight commanded by Valery Bykovsky, which would involve crewmembers Alexei Yeliseyev and Yevgeni Khrunov transferring to Komarov's Soyuz by a spacewalk. However, Komarov's flight was plagued by serious technical issues and the second flight was abandoned. Komarov used every ounce of skill he had to pilot his capsule back into the Earth's atmosphere but on re-entry the parachute system became caught up with the module, and, with no

Pravda newspaper announces Komarov's death, 25 April 1967. (*Author's collection*)

means of escape (ejection options had been removed from the Soyuz craft), plunged into Earth and caught fire, killing Komarov. There were recriminations and Gagarin and Leonov insisted that Kamanin record their concerns in the official investigation about the lack of attention to the previously identified faults with Soyuz – particularly criticizing the new chief designer Mishin.

Work on the lunar programme continued despite all the setbacks but the plan for a manned orbit of the moon in May 1967 now began to slip out of sight, and a moon landing seemed a quite distant prospect. The next manned Soviet spaceflight, Soyuz 3, did not take place until October 1968. This was piloted by Georgi Beregovoi, at 47 years old the oldest of the cosmonaut team and a second-generation trainee. Beregovoi was a very experienced test pilot and, as a Sturmovik ground-attack aircraft pilot under Kamanin's command in the Great Patriotic War, had won an HSU award for bravery. His mission was to dock with the previously launched unmanned Soyuz 2 while orbiting the Earth, a significant achievement but conservative in comparison to the pace of the space race by this time. Beregovoi met up with Soyuz 4 but failed to achieve docking and returned to Earth safely. Officially published photographs of Soyuz 3 at its launch were the first showing the Baikonur site to the Soviet public and the world.

The Americans meantime were leaping ahead in their efforts to carry out Kennedy's goal of a man on the moon by the end of the decade. NASA's Gemini Programme of 1965/6 had involved ten successive manned missions in low Earth orbit that had provided a sound base for continued progress with the more ambitious Apollo programme, initiated in 1962. This too had setbacks, not least of which was the destruction by fire of a cabin with a three-man crew aboard during a pre-launch test in 1967. On 21 December 1968, the Americans achieved the first manned orbit of another planet when Apollo 8 with its three-man crew successfully went around the moon and returned to Earth. On 21 July 1969 Neil Armstrong set foot on the moon after Apollo 11's landing, a momentous occasion that concluded the race with the Americans as clear winners.

The Soviets tried to create the impression that they had lost interest in the race to the moon, but their continued efforts to develop the N-1 rocket suggest otherwise. The scrap from the cancellation of the project in 1974 remained hidden until after the Soviet era had ended. Mishin wanted to use some aspects of N-1 and the lunar programme development to build an ambitious moon base – the Zvezda programme that had been in the planning stages since 1962. His successor in 1974, Glushko, tried to continue, but lack of investment in the rocketry required after the failure of the N-1, meant that it never made much progress. Moon exploration within the limits of available technology remained an active part of the Soviet space programme until emphasis and investment shifted to space station development. The American Apollo 11 mission involved a moon vehicle that accompanied the crew and gathered samples. The Soviets had attempted to land a robotic moon vehicle, Lunokhod, but the launch in February 1969 exploded upon lift-off, resulting, it was later revealed, in the spread of a large amount of radioactive material across the region. All this remained secret, but the successful landing on the moon of a Lunokhod rover in November 1970, which sent pictures and soil analysis back to Earth, was celebrated as a Soviet triumph. A more sophisticated Lunokhod rover undertook a further successful mission in January 1973, but a third mission, planned for 1977, was cancelled. Lunokhod 1 continued in operation for 322 days, travelled about six miles and sent back thousands of films and photographs, as well as other scientific information.

Its successor, Lunokhod 2, travelled over rougher terrain for over 40 miles for four months, returning even more information and images. Both are still on the moon, landed objects among the more than estimated twenty-eight million pieces of manmade detritus left in outer space since the first penetration of the Earth's atmosphere in 1957 (Natural History Museum 2019).

The Death of Gagarin

Soviet self-confidence received another body blow on 27 March 1968 with the death of Yuri Gagarin. As early as mid-1962 Korolev had been complaining about the effective loss of two key cosmonauts to his programme: Gagarin and Titov. In their ambassadorial roles, both were falling behind with the required academic study and fitness training. In Gagarin's case, Korolev succeeded in bringing him back into the fold when he was appointed, already an air force colonel, as deputy head of the Cosmonaut Training Centre in December 1963, second only to Kamanin.

Gagarin was always keen to recommence cosmonaut training but, as the first man in space, was too precious a commodity to risk. He continued to play an important role in the background of the space programme, but the deaths of Korolev and Kamarov affected him badly and, according to Kamanin, he reverted to self-destruction through partying and drinking. All the while he continued his academic education, as did the other cosmonauts, and eventually graduated from the Zhukovsky Air Force Academy in February 1968 having successfully presented a thesis on the subject of spacecraft aerodynamic configuration.

Gagarin remained keen to retain his pilot skills even though he knew he might never return to space, but flying time was maintained at a low level for cosmonauts, unlike their American counterparts.

Between 1960 and 1968 Gagarin accumulated less than ten hours' flying a year, and none of it solo. It was in an attempt to redress this, redirect his life and win back the respect of the other cosmonauts, that he made an effort to catch up on his flying hours in early 1968. At this time his wife and daughter were living in his home town of Gzhatsk, while their new apartment in Star City was awaiting completion. He was scheduled to visit them in early December 1967. Gagarin finally took to the sky again on 27 March in a tandem MiG-15 trainer with an experienced pilot, Vladimir Seregin. It was one of his first flights in five months. To this day the facts surrounding their fatal accident are disputed. Exactly how and why their aircraft crashed into the ground without giving them time to eject is not known. The official investigation, whose findings were kept secret for many years, concluded that it was pilot error, but a KGB report released in 2003 revealed that the pilots had been given incorrect information by ground control. Later theories by those familiar with the aircraft suggest it was a mechanical fault, and, most recently, Leonov, one of the first on the crash scene, has stated his belief that the MiG-15 spun out of control while avoiding another jet fighter that had accidentally crossed their flight path. Conspiracy theories also abound, their contentions ranging from assassination by jealous Soviet rivals (including Brezhnev) or the American CIA, to a collision with a UFO. Others contend that he remained alive after the crash and was hidden in a psychiatric hospital until his real death in 1990. These theories are summarized in an *Independent* newspaper article of 28 July 2005.

Valentina Tereshkova comforts Valentina Gagarin at Yuri Gagarin's funeral, 1968. (*TASS/Musaelyan Vladimir; Savostyanov Vladimir, Alamy Stock*)

Valya was in a Moscow hospital at the time of the accident, undergoing an appendix removal and Gagarin had planned to visit her that evening. She now had to contend with his loss and the fact that, as he had died in a massive explosion, she would not even get to see his body. For the Soviet nation, Gagarin's death was of immensely greater significance than that of Korolev (whose name had been kept secret) and Komarov. The nation went into mourning for their best-known citizen whose smiling, friendly face was known throughout every corner of the world. The funeral, ending with his burial in Moscow's Kremlin Wall, was a massive event with some in the crowd voicing disbelief that such a young man could die so pointlessly. Funerals were now replacing the celebrations of previous years that had greeted the successful heroes on their entry to Red Square. In April 1961 Gagarin left a letter for his family in the event of his death during his pioneering first space flight. Given to Valentina Gagarin after his death in 1968, it was not released to the public until 2011 (fifty years after his flight). It reads:

Hello, my sweet and much loved Valechka, Lenochka and Galochka. Here I've decided to write you a few lines to share the joy and happiness I felt today. Today a governmental commission decided to send me first to space. You know, dear Valyusha, I'm so happy; I want you to be happy with me. A simple man has been trusted such a big national task – to blaze the trail into space! Is there anything bigger to wish for? This is history, a new age! The day after tomorrow is the launch. You'll be doing your regular things then. It's a very big task lying on my shoulders. I wish I had a chance to be with you for a little while before it, to talk to you. Alas, you are far away. Nevertheless, I always feel you by my side. I trust the hardware completely. It will not fail. But it happens that a man falls right on a level ground and breaks his neck. Some accident may happen here too. I personally don't believe it would happen. But if it does, I ask you all and you, Valyusha, in the first place, not to waste yourself with grief. Life is life, and nobody is safe from being run over by a car. Take care of our girls; love them like I do. Please, raise them not as some lazy mommy's girls, but real persons who can handle anything life throws at them. Make them worthy of the new society – communism. The state will help you do it. As for your personal life, settle it the way your heart tells you, the way you feel right. I hold no obligation from you and I don't think I have a right for it. This letter seems too gloomy. I don't feel like it. I hope you'll never see this letter, and I'll never have to be ashamed for this moment of weakness of mine. But if something goes wrong, you have to know it all. So far, I lived an honest, rightful life; I served the people, even though this service was a little one. In my childhood I read Valery Chkalov's legendary words: 'If being, then be first'. Well, I'm trying to be one and I will be to the end. I want, Valechka, to dedicate my flight to the people of the new society, communism, which we are about to become part of, to our great motherland, to our science. I hope in a few days we will be together again and will be happy. Valechka, please, don't forget my parents, and if you have an opportunity, give them a helping hand. Send them my biggest greetings and ask their forgiveness for my keeping them unaware; they are not supposed to know anything. This seems to be all. Goodbye, my dears. I embrace you all tight and kiss you, your dad and Yura. 10 April, 1961. Gagarin.

(RT World News, 30 March 2011)

Valentina Gagarin moved back to Star City with her children and was given a pension and accommodation – the newly completed apartment the family had been allocated shortly before her husband's death. She continued to work out of the public eye in the Star City Health Centre until she retired, refusing press interviews but involved in promoting the life and work of her late husband. She died in Star City in March 2020 at the age of 84. The couple's daughters were both successful with their chosen careers as adults and both had children of their own. Gagarin's name lives on throughout the world – streets all over have been named after him and his place in human history is assured despite the demise of the Soviet state that produced him and which he served faithfully.

From Soyuz to Salyut, and Beyond to Venus

Although the dream of being the first on the moon had now been transcended by American success, there were still other attainable 'firsts' to be sought by the comparatively underfunded Soviet programme. The modular design of the Soyuz craft lent itself to space station development, and the mission that had ended in Komarov's death was eventually carried out successfully in January 1969 by Vladimir Shalatov in Soyuz 4, and Boris Volynov, Alexei Yeliseyev and Yevgeny Khrunov in Soyuz 5. The two vessels docked and Yeliseyev and Khrunov walked in space for thirty-seven minutes between the two craft. They then separated and continued their mission. Soyuz 4 landed successfully with its (now) crew of three, and Volnyov landed in Soyuz 5 the next day. In October 1969 experiments were extended to three Soyuz craft (Soyuz 6, 7 and 8) with seven astronauts between them, which manoeuvred close to one another, with each spending five days in space. The next flight, of Soyuz 9, in June 1970, demonstrated that prolonged periods could be spent under weightless conditions in space, with a stay of eighteen days in orbit of the Earth. In the meantime, development of a space station, Salyut (Salute), which had started under Korolev, had continued and was launched into an Earth orbit in April 1971. A few days later Soyuz 11, with three cosmonauts aboard, arrived alongside and successfully conducted preliminary linking and docking. Soyuz 11 returned to Earth while Salyut continued with its automated orbit. This progress continued with the successful joining of Soyuz 11 and Salyut on 6 June 1971. The

The Race to the Moon: Failure and Retreat   147

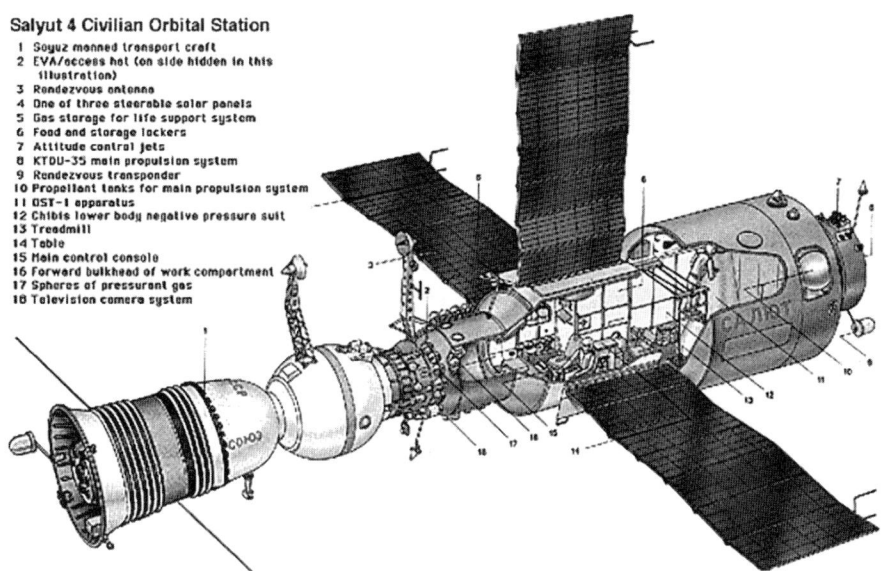

Diagram of space station Salyut and Soyuz spacecraft. (*NASA via Wikimedia Commons*)

crew then moved into the now-extended space station, consisting of the two craft, and remained there for twenty-three days. The mission ended in disaster, though, when the planned re-entry by Soyuz 11 to Earth went wrong and the capsule became de-pressurized, killing the crew of Georgi Dobrovolsky, Vladimir Volkov and Victor Patsaev. Although Salyut's purpose was ostensibly peaceful it was used as a cover for the highly secret military Almaz programme linked to the Salyut 2, 3 and 5 launches undertaken between 1973 and 1976, as a response to American military space developments. These three Salyut craft were all defensively fitted with a cannon – the only examples of armed spacecraft since the advent of spaceflight. Salyut was succeeded by Mir (Peace) in 1986, which operated successfully until after the Soviet era.

Meanwhile, unmanned satellite missions continued – including missions to explore Venus between 1961 and 1984 (the Venera Programme), and Mars between 1960 and 1973 (the Mars Programme). Most investments in the later years of the Soviet Union went into space station development, involving a degree of international cooperation and the sharing of technology. These were largely successful and their use and development continue to this day, albeit as commercial operations in Russia since 1991 (discussed in Chapter 9).

* * *

The beginning of the inevitable end of the expensive space race started in 1971 with preliminary talks between the Americans and Soviets on joint missions linking Apollo and Soyuz spacecraft. These eventually came to fruition in July 1975 with the successful docking in space of the two craft and a symbolic handshake between the commanders Alexei Leonov and Thomas Stafford. The Cold War would continue in the years to come, but the space race was over. The Soviets had been overtaken by the power of the dollar: a secret CIA document (released to the public in 1998) shows that in 1957 the USSR was spending between $0.2 and $0.3 billion on their programme, compared with $0.1 billion by the US. By 1964 American spending had jumped to over $6 billion while the Soviet's was thought to be between $2 and $4 billion (CIA August 1964). After the end of the Khrushchev years, when support began to waver, that spending gap widened.

The people of the Soviet Union had lost interest in space exploration by the early 1970s. Many were concerned with their own problems. The technology involved in space exploration seemed a waste now it no longer

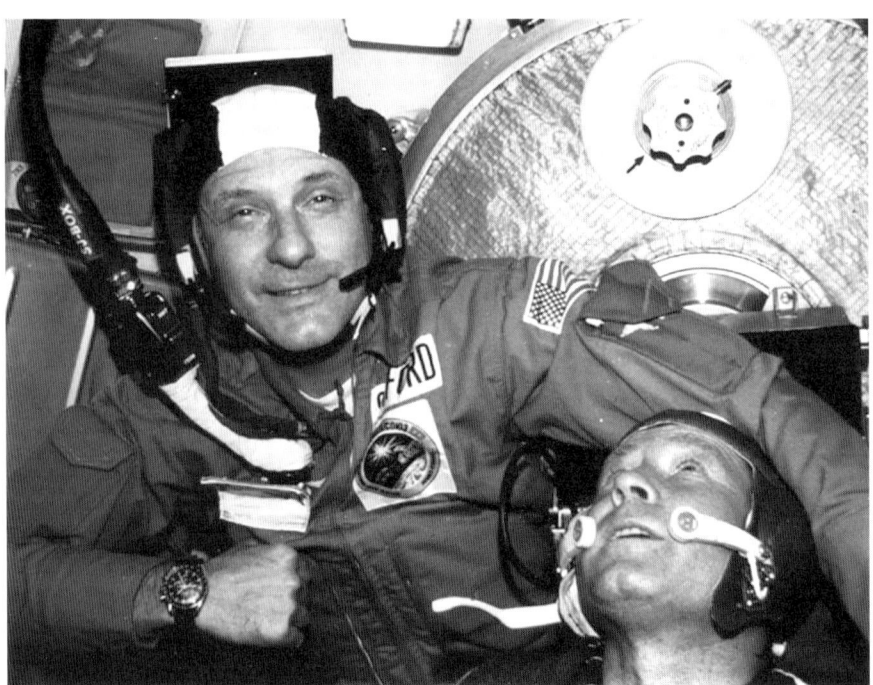

Cosmonaut Leonov and astronaut Stafford meet in space, 17 July 1975 (*NASA via Wikimedia Commons*)

involved the pride in achievement experienced ten years earlier. A new generation was emerging who had no memory of the war, its sacrifices and the patriotism it called upon; they wanted a better life for themselves, on Earth, and were no longer dreaming of the cosmos.

The Americans may have won the race to the moon, but some involved, as NASA scientists do not, with hindsight, regard their successes as total. A 2016 film documentary *The Fight for Space*, which gives voice to several former NASA scientists, is clear that the US space programme was all about beating the Soviets and included little scientific or exploratory purpose – despite claims to the contrary at the time. Until the Apollo 17 mission to the moon in 1972, all astronauts were air force pilots – unlike the Soviets, who included scientists and engineers from as early as Voskhod-1 in 1964. Once the Cold War was over, American governments rapidly lost interest in space exploration, and developments from then on were comparatively modest and lacking in purpose. Admiration by these NASA scientists was expressed for the Soviet Vostok space capsule, for which they never had an equivalent. The Vostok formed a solid foundation which continued through the development of Voskhod to the Soyuz, which is still in use over fifty years after its debut.

Chapter 8

Building Communism on Earth: The Virgin Lands Campaign and the Baikal–Amur Railway

There will be no communism if our country has as much metal and cement as you like but meat and grain are in short supply.
                                    Nikita Khrushchev, 1954

The space programme was the ultimate prestige project in the Soviet Union – taking communist ideas beyond the boundaries of our planet and opening up new worlds. However, from the time of the revolution in 1917, the Bolshevik leaders, and their Communist Party heirs, realized that dreams of the future were one thing, but humans needed to feel things were improving in the here and now to help motivate them to build a brotherhood of mankind and a new world of socialism on the road to eventual communism. The enthusiastic participation of the people in creating their future was a tenet of Bolshevism. Lenin's call, which resonated with the war-weary masses in 1917, was for 'Peace, Land and Bread' and he invited them to seize these things for themselves. In 1920, he declared that communism would be built on electrification of the whole country. By the 1950s and 1960s, the people of the Soviet Union were looking for more than essentials, but the country remained well behind the advanced Western nations in terms of rewards offered to the people in return for their toil. What it did provide was employment and freedom from worry about basics like education and social security. However, there was a bureaucratic uniformity attached to these achievements, and the available consumer items, that made the country seem drab and dull. There was still much to do to bring up living standards and even electricity remained a dream for many in rural areas. While the space successes demonstrated that the Soviet state was capable of the most advanced forms of science and engineering, the reality felt different. Stalin had been the architect of the Five Year Plans and the rapid efforts to industrialize the country in the 1930s, and both Khrushchev and Brezhnev had been senior Party officials responsible for realizing Stalin's dreams. His style, if not all his ruthless methods, had rubbed off on both.

On the notion that people will endure hardship, if they are motivated by a clear goal that makes their sacrifices seem worthwhile (as had been the case in wartime), post-war regimes sought to improve economic progress and engage people, especially the young, in prestige development projects. The two most significant in terms of investment, both human and financial, will be discussed in this chapter: the Virgin Lands campaign of the 1950s and the building of the Baikal–Amur Railway (BAM) in the 1970s and 1980s. The space programme represented the cutting edge of progress towards communism, but whether these Earth-based projects lived up to the same ideals, and achieved anything lasting, was quite another matter.

The Virgin Lands Campaign

When he took over as Soviet leader after Stalin's death in 1953, Nikita Khrushchev inherited a problem that had plagued the USSR since the 1917 revolution: how to feed its people. Stalin planned to build socialism in the 1920s and 1930s through rapid industrial growth and increased food production to feed the new urbanized workforce by way of forced collectivization in the countryside. The latter was unpopular to the point of active resistance by the wealthier peasants (the *kulaks*) and inefficiencies resulted in only partial success in increasing agricultural production. The fundamental problems across the vast expanses of the USSR were related to geography and climate. The periodic droughts and famines that had always affected the countryside long pre-dated the Soviet era (Golubev & Gronin 2004). Some issues around these matters should be clarified at this point. Anne Applebaum, a popular contemporary historian, has only focused on the human political contribution to such disaster (*Red Famine* 2017) when describing the Ukrainian famine of 1932/3 which killed millions; however, other natural factors had played a major role. The general conclusion of experts, concerning acute food shortages across the world, is that drought and the disruption it brings to the local food supply are natural, but famines have human and political causes and are generally avoidable.

Comparison with the US is instructive: during the Soviet era, both countries were about equal in land mass (8.5 million square miles) and population (about 200 million). Both had agricultural heartlands surrounded by wastelands where almost nothing could grow, but the Soviet Union's arable land was much smaller, at 386 million acres compared to the USA's 440 million. However, it is the Soviet Union's northern latitude that is crucial: two-thirds of the country consists of taiga (dense northern coniferous forests) and Arctic tundra, compared to less than half of North America's landmass. Permafrost affects about half of the Soviet Union, and the northern fertile regions have short summers and short growing seasons. The southern areas, which enjoy higher temperatures, are dry and prone to drought, with the regular absence of rain. The further east one goes in the areas of the old USSR, the dryer the climate becomes due of the absence of precipitation from nearby warmer oceans – Siberia would be mostly desert were it not so cold. The consequence of all this is that in the 1950s, Soviet yields of wheat per acre were about twelve bushels, while in the US seventeen bushels was the norm; for corn, the comparable figures are seventeen bushels to forty bushels (Harris 1955). The Soviet Union's disadvantage is clear even before any consideration is given to the very different systems of land ownership and agricultural production.

Veniamin Nikitsky from Kharkov, Ukraine, in his teens in the mid-1950s, recalls the general shortage of food at this time. Bread and wheat products disappeared from the shops and the family lived on soup and porridge, on produce from their garden in the summer, along with hens and rabbits kept for food. This was sometimes supplemented by supplies his father secured in work-related travel to the far east of the country where fish and even caviar were occasionally available.

Stalin had paid little attention post war to the problems of food production, so Khrushchev realized he could make his mark by addressing the issue. The harvest of 1953 had been a poor one and even an increase of 10 per cent would not have been enough to feed the cities given the rapid scale of urbanization at that time. To achieve the same levels of food output, taken for granted in the USA and Western Europe, would require a doubling of the production of grain, tripling that of meat and

The Virgin Lands Campaign and the Baikal–Amur Railway 153

even greater increases for poultry and eggs. Looking at the map it was clear that there were warmer areas in the south and southeast of the country that could be developed for agriculture. These were in remote and sparsely populated regions where they had been typically used, if at all, for non-intensive livestock grazing. Problems of low rainfall and drought would be ignored in a rush to resolve the USSR's food supply problem by increasing the grain yield. This, it was thought, could even turn the country into a grain-exporting nation like the USA, and enhance prestige and influence with the developing countries. Thus was born the Virgin Lands Campaign.

Khrushchev's big idea was to open up such underused and fallow land in north Kazakhstan, Western Siberia and south-eastern Russia. Ukraine, the traditional breadbasket of the USSR, would be turned over to maize production for cattle feed so that stimulus could be given to the underproductive meat and dairy sectors. All this would be a stop-gap measure pending increases in agricultural production through new fertilizers, technical development and mechanization. This would be a popular campaign with mass involvement that would inspire and motivate the new generation of young Soviet citizens. There was opposition from the Kazakh leadership, arising from genuine worry about the environmental impact of the proposals (including the displacement of traditional livestock herders), and from concerns about the influx of Russians to a republic that already considered itself a minority suffering some degree of discrimination. Khrushchev's response was to send Brezhnev to persuade them of their duty to the nation and bring them into line. By 1956 he was the first secretary of the Kazakh Communist Party. Matters moved quickly, the new agricultural policy was passed at the Central Committee of the CPSU in January 1954 and by the newly invigorated Kazakh Party the next month. Brezhnev then spent a considerable time ensuring that the policy was put into effect on the ground, to the point of becoming involved at a very local level when it was considered necessary to sort out recalcitrant officials and others who stood in the way of 'progress'.

By 1955, eighteen million hectares of land had been ploughed and prepared in the selected areas, and by the end of the campaign, this had reached forty-two million. This included 96,526 square miles of steppe land in northern Kazakhstan – an area larger than the whole of Britain. This extraordinary achievement involved the deployment of hundreds of

Montage of images from Brezhnev's book on the Virgin Lands Campaign. Clockwise from top left: Brezhnev in 1956, 'Komsomol members leave for the Virgin Lands', 'Bumper Crop', 'cooking breakfast on outdoor stoves at Slavyansky State farm'. (The Virgin Lands, *1978*)

thousands of Soviet citizens to the areas involved. In the summer of 1954 alone 300,000 Komsomol volunteers set off from the cities and towns of the western USSR to the Virgin Land regions, many leaving to great fanfare in special trains. While many were young Komsomol members whose contractual periods were short, whole families also moved to make new lives for themselves.

New settlements were built from scratch to house the settlers – starting with tented cities that gradually became more permanent as the work on the land proceeded apace. The urgent requirement to produce more food precluded any notion that the infrastructure to support the campaign might have come first. This involved great hardship for those involved – temperatures could rise to extreme heat during the day and sink to sub-zero at night. Many (and this is admitted in Brezhnev's account) found this too much and returned west. However, new heroes were celebrated with awards and references in Soviet literature and newsreels of the period. Brezhnev (1978) mentions a number including Leonid Kartauzov (1933–2016), a tractor and combine harvester driver from Leningrad working in northern Kazakhstan whose efforts were inspirational to others. Kartauzov had lost both legs while serving as a child soldier during the war and used prostheses. He was recognized as a Hero of Socialist Labour in 1972 and his story was made into a short film, *A Ballad About a Human Being*, in 1978. As a 22-year-old he had travelled to the Kostanay region in March 1955 having learned of the Virgin Lands Campaign from relatives. He was apparently obsessed with a desire to feed people after his father had died from hunger during the siege of Leningrad in 1941. Starting in his trade as a shoemaker, he succeeded

Virgin Lands postcard set from 1955. (*Author's collection*)

Leonid Kartauzov. (The Virgin Lands, *1978*)

in persuading managers that he could work as a driver and gradually achieved qualifications and became an exemplary worker.

Hundreds of new collective farms (*kolkhozes*) were created to manage the tens of thousands of hectares of land allocated to each one. The centralized planning behind this also ensured that almost all newly produced agricultural machinery – tractors, combine harvesters and other technical equipment – went to the Virgin Lands regions. Up until 1958 machinery was organized centrally through Machine and Tractor Stations (MTSs), but this was changed so that each collective farm was directly allocated what they might require. The names of the kolkhozes often reflected the cities from which the groups who built them had come: Moscow, Minsk, Kiev, Dnepropretrovsk, Aramarvir, Tagil, Sochi, Poltava, Yaroslavl, etc. These new settlements lacked a balance that would make them secure for the future as they were initially often populated by young men seeking adventure whose planned stay would be short, partly because there were few women with whom to potentially settle into family life. The state newspaper *Pravda* printed a letter from a group of women from the Marinovka Kolkhoz, Atbasar District, northern Kazakhstan on 17 July 17 1954, appealing to young women to come to join them in the Virgin Lands. This received a good response and the new influx of women helped transform the pioneering settlements into

The Virgin Lands Campaign and the Baikal–Amur Railway 157

more conventional and sustainable communities. As was typical in the USSR, equality of the sexes was more aspirational than actual – women tended to work in traditionally gendered roles as milkmaids, cooks and in secretarial jobs, rather than working the machines out in the fields – although there were exceptions. Their situation generally in kolkhozes at this time was dire; a woman's eligibility for marriage was measured by her strength, her labour day credits (a value put on work for the collective farm), and her dowry – charm, intelligence and compatibility came way down the list (Belov 1956).

As the settlements increased in size, the infrastructure was put in place to sustain them; of vital importance was the supply of electricity. Alexander Kudinov, whom we met in Chapter 5, was, by 1966, a Komsomol member studying electronics at college in Kharkov. He spent the summer of 1967 with a student construction brigade erecting electric supply lines in Tyumen, Western Siberia. He spent the following summer of 1968

Farmworkers at the Malenkhov Kolkhoz, Gorky region, hear news of the 1954 agricultural plan. (Soviet Union *magazine, July 1954*)

158   Soviets in Space

Posed image of young peasants courting, Stavropol. (Soviet Union *magazine, September 1953*)

in the Kostanay district of northern Kazakhstan building a mechanized threshing floor in a Virgin Lands kolkhoz. This, like others, was inhabited purely by Russians and Ukrainians who had been developing the area for over ten years by the time of his stay. In just two months of hard but well-paid work, Alexander received the same as eighteen months' worth of student allowance. He remembers enjoying these experiences – believing at the time that he was contributing to the well-being of the nation.

All those who spent two years or more as participants in the Virgin Lands Campaign received a medal – over one million were awarded. This was also given to Yuri Gagarin and after that, as a matter of custom, to the cosmonauts who followed, in symbolic recognition of their opening up of the Virgin Lands of outer space. Pin badges were also given to Komsomol members for their contribution to the meeting of annual quotas.

The results of the campaign were initially spectacular. By 1956 grain production had increased by three times on that of 1953. A portent of future problems should have been seen by the poor yield for the very dry year of 1955, but 1956 was exceptionally good in the Virgin Lands with good rainfall leading to a bumper crop. This came in the same year the

The Virgin Lands Campaign and the Baikal–Amur Railway 159

Virgin Lands pin badge for helping meet quota, and two years of service medal. (*Author's collection*)

western USSR suffered a drought, so the overall result was a success and famine and catastrophe were avoided. Even so, not all the wheat could be harvested before winter due to the shortage of tractors and mechanized equipment (and railway wagons for transport of what was timeously harvested). Thus, some was lost as a result of success rather than failure. By 1962 grain output had increased, with some annual fluctuation, from 85 million metric tons in 1954, to 140 million in 1962. This was mirrored in other sectors such as meat, milk and egg production (Strauss 1969).

Khrushchev's reputation was saved and his popularity assured. However, he had emphasized rapid production increases and had ignored the basic science of crop rotation. Year-on-year exploitation of land inevitably results in soil erosion and this is exactly what happened. The Seven Year Plan, announced in 1959, was based entirely on continued increases in output and proved totally unrealistic. By the early 1960s, a series of windstorms across the steppe had blown away much of the remaining fertile topsoil which led to a vastly reduced area for cultivation. This resulted in the loss

Combine harvesters, Krasnodar region, 1961. (*TASS/Alamy Stock*)

of almost half of the Virgin Lands by 1965. Associated developments within the Five Year Plan to grow maize and use it to increase livestock farming were similarly successful in the short term. Again, the absence of crop rotation and the reduction of the traditional, but essential, hay crop resulted in a rapid decline in overall production. Livestock targets were overestimated so that in some places maintenance of the annual quota resulted in the slaughter of dairy as well as beef cattle – creating problems for future years. In the Ryazan district the party secretary,

Delivery of grain from the Udarnik State Farm, Samarkand, Uzbek SSR. (Soviet Union, *July 1954*)

Larionov, did exactly that: exceeded the quota and was made a Hero of Socialist Labour in 1959. He shot himself when everything collapsed the following year. The change away from MTS responsibility for machinery was not smooth – those with an interest in the old system placed obstacles in the way, and kolkhozes found themselves in debt, owing large amounts of money for the assets they were allocated and were now responsible for.

The production increases that were achieved were not enough to sustain the needs of the population. By 1962, food shortages were such that price increases were announced, causing serious riots in Novocherkassk and unrest elsewhere. Khrushchev's popularity was declining despite the space successes and the fact that he had at least tried to address food production issues. The matter got worse in 1963 through drought and a bad harvest. Khrushchev was reluctantly forced to spend a proportion of the USSR's cash reserves on imported grain and other foodstuffs. In 1964, this, and other issues, caused his un-ceremonial removal while on holiday and forced him into an unhappy retirement.

Centrally controlled planning, as exercised in the USSR, had many virtues; the mass relocation of industries in 1941 from areas about to be occupied by the invading Germans, to areas far to the east, saved the country. However, how Khrushchev used central economic planning to launch successive popular campaigns that had a poor scientific basis typifies the worst aspects. The areas of the Virgin Lands Campaign suffered from poor and erratic rainfall and, in later years, other plans were considered to reverse and reroute rivers to provide irrigation to arid regions of the USSR. These were a continuation of Stalin's Great Plan to

Transform Nature of 1948 that planned irrigation canals and windbreak tree plantations that would bring life to barren regions of Kazakhstan. That too never lived up to the dreams of the planners. Some of the Brezhnev plans to divert rivers from (in the planners' view) uselessly flowing into the frozen northern Arctic would have involved a series of small nuclear detonations with even more catastrophic potential than the intended outcome. The river reversals thankfully never took place but other plans were eventually, if unsuccessfully, realized: notably, to irrigate areas for cotton plantation in Uzbekistan, resulting in the draining and decimation of the Aral Sea – a lasting legacy of environmental disaster. Ironically, the USSR remained dependent on the USA and other Western countries for grain and other food supplies.

The Construction of the Baikal–Amur Railway (BAM)

While the Soviet Union lacked rainfall sufficient to enable it to produce enough food consistently, it had an abundance of other useful natural resources. Brezhnev's plan to open some of these up in the remotest Siberian regions of the country was an attempt to make the same mark on history as Khrushchev had with his space programme and Virgin Lands Campaign. The construction project was proclaimed as a major component of the tenth Five Year Plan at the 25th Congress of the CPSU in 1976. However, the initial development work had started several years earlier. Brezhnev's intent to make this a prestige project that would capture the imagination and voluntary involvement of citizens was summed up in the slogan that the railway would be made 'by clean hands only'. This was a direct reference to the fact that previous giant construction projects of the Stalin era had relied on plentiful supplies of prison labour, but his would be created by the enthusiastic hands of the people. Stalin had announced to the world in 1936 that socialism had been achieved; Khrushchev followed this up in 1961 with his commitment to achieving the next goal of communism by 1980. While this seemed an impossibility in the 1970s, the commitment to continue to build towards communism remained. The BAM represented a tangible sign of practical ('developed') socialism (terms used by Brezhnev and the planners from the 1960s onwards) leading the country in the right direction, even if the deadline was unattainable.

The proposal was to build a railway parallel to the existing Trans-Siberian line, that would exit the existing route at Tayshet in Siberia and then run north of Lake Baikal to finally reach the Pacific coast at Sovetskaya Gavan – a total distance of 2,687 miles through largely unpopulated areas, difficult terrain, and some of the wildest and most inhospitable places on Earth. A start had been made using prison labour in the 1930s and 1940s. A line already existed between Tayshet and Bratsk – a distance of over 200 miles – and at the other end from Sovetskaya Gavan to a point near Komsomolsk, some 300 miles eastward. Tynda, approximately midway, had already been reached by a 110-mile branch line from the main Trans-Siberian Railway to the south. The aim would be to open up the interior for population expansion with the creation of new towns and settlements along the railway, and the associated exploitation of mineral and logging resources. Huge investment was promised to fulfil the plan – 1 per cent of the USSR's GDP in the years from 1974 to 1984, between $15 and $20 billion (Ward 2009). This enormous sum would build thousands of bridges to cross expansive and fast-flowing rivers, cut through impenetrable forest, and move some 300 million cubic feet of earth, much of it from the dozen or so long tunnels that would be required.

Brezhnev's plan to construct BAM involved the mass mobilization of young people through the Komsomol, whose organs would be given responsibility for the project – effectively managing the array of volunteer labourers as well as technical experts and support services in the new

Route of the Baikal–Amur Railway. (*openstreetmap.org via Wikimedia Commons*)

settlements along the line. Much was at stake, the regime had come in for criticism for the continued crushing of growing dissent both at home and within the satellite countries – the Prague Spring of 1968 and its violent aftermath as Soviet tanks invaded the country and brought its leaders to heel, being the best-known example. Young people were becoming fractious in their envy of Western culture, and, hooliganism, was increasing. Women were frustrated at being expected to work and undertake their traditional domestic role, and many ethnic minorities and Soviet nationalities were raising voices of discontent. Those who participated in BAM could provide the base for the USSR's new generation of model citizens, embodying the pioneering spirit and loyalty to the state that the cosmonauts had exemplified in the 1960s – this time through mass activity. BAM had the potential, or so the planners hoped, to unite the nation in a project that could be built with their own hands, and which might bring the economic benefits they so longed for. BAM would not only open up potentially rich areas of natural resources but could also be used to export Soviet raw materials to the growing capitalist economies of Japan and South East Asia in return for the superior consumer goods so sought after in the USSR.

Komsomol volunteers leaving Moscow for BAM construction work, 25 September 1974. (*E. Kotliakov RIAN via Wikimedia Commons*)

The Virgin Lands Campaign and the Baikal–Amur Railway

Montage of images from 1977 Soviet book about BAM. Clockwise from top left: 'The first section is being laid', 'Faster, still faster!', 'I'm going to carry you through life just like this', 'What are you frowning at, you've managed above quota, haven't you?' (*The Great Baikal–Amur Railway*)

So much for the dream. What of the reality? The BAM project certainly captured the imagination of hundreds of thousands of the nation's youth. By the end of 1974, over 50,000 of the 156,000 young people who had applied to help build the railway, had arrived in the remote areas where construction was taking place. By 1984, half a million young people had been involved for varying lengths of time, two-thirds of whom were Komsomol members. Most were motivated by the ideals behind the project, some by high wages and the opportunity for adventure, but there were also hardened criminal types seeking new opportunities away from the rule of law. The volunteers came from all over the Soviet republics and beyond – some from the satellite countries of Eastern Europe and friendly nations such as Cuba, Mozambique and Angola. From the outset there was more balance between the sexes than had been the case in the Virgin Lands – 40 per cent were women, half of whom were under 20 years of age. The project also involved professional railway workers and conscript soldiers (who made up an estimated 25 per cent of the workforce), a tough assignment, but better than service in Afghanistan.

Most of those who volunteered for the project had no training or technical skills and their enthusiasm could hardly make up for this.

Photographs from a BAM soldier constructor's album, 1981/2. (*Author's collection*)

A Yakut Komsomol construction team, BAM, 1977. (*V. Yakovlev RIAN – Wikimedia Commons*)

Inevitably this resulted in waste and, from the outset, failure to reach construction targets in this very harsh environment. By 1976, over a hundred miles of track had been laid. This would have had to be tripled to achieve Brezhnev's promise that the railway would be built by 1984. Although the railway was declared complete that year when construction gangs from east and west met in a fairly quiet ceremony, in reality, there was no through traffic until 1991 when the line was again declared complete, although operational on only a limited basis.

The Soviet Union's artists' and writers' union members were mobilized to build the public face of BAM. These were the 'engineers of the human soul' as Stalin described them, and their role was considered important in growing the beliefs and ideals that lay behind the project. This was far more than just imaginative expression of a vast construction – like the building of the great Boulder and Hoover dams and other public projects that helped the US out of the depression of the 1930s, celebrated through the officially commissioned songs of Woody Guthrie – as BAM was to reflect all that was good about the USSR and would be a source of pride for its citizens. Their contributions filled the state press and other publications of the era. Yevgeny Yevtushenko, the Soviet Union's best-known living poet wrote a long and complex poem, *The Forest Glade*, dedicated to the *bamovtsy* (men) and *bamovki* (women), as the constructors were collectively known. In this excerpt, he colourfully describes an incident he witnessed in 1974:

> I'm on a ferry –
> Like on the very
> Brink of the river's other side.
> Mud spattered lorries,
> Quiet, in no hurry,
> Gaze at the Lena – majestic, wide.
> A lorry driver
> Reads about China,
> While BAM girls twitter, aged seventeen,
> Discussing pullies, pantyhose and woollies,
> which here, at BAM-shops aren't to be seen
> An excavator
> Left by some blighter
> Submerged in mid-stream up to the waist,
> Stuck in a river,
> Pities, a-shiver,
> The Japs who made it not just for waste.
> A timber-feller,
> Huge-whiskered fellow,
> Grins, looking at it beneath his hand:
> 'It's my assumption,
> Those folk had gumption
> Who chucked in mid-stream
> A hundred grand!
> And all in dollars!
> To think such fellers
> Can still be met in the Soviet land!'
>
> (*The Great Baikal–Amur Railway*, 1977, p105)

The poem goes on to describe the rescue of this expensive but discarded piece of equipment through the spontaneous organization of the ordinary worker – Yevtushenko's tribute to the socialist commitment and spirit of BAM's workforce.

A piece, by journalist Yuri Balakirev, in the same propagandist publication, described some of the settlements and workers he encountered on the line. At the lineside community, known originally as Kilometre 85 but renamed Anosovskaya, he met Lida Syratavtskaya, a teacher with a class of five pupils in a new school that had opened the previous year, 1974. The streets had names such as Trailblazer's and Komsomol, and already included a general shop, a bakery, a canteen and two childcare centres. Since the community opened up there had been thirty-eight weddings between young Komsomol members, the latest between tractor driver Vasya Cherny and technician Lena Todorova. Thirteen babies had been born in Anosovskaya and the assumption was that they, and their parents, would become long-term inhabitants and that the completion of the railway would bring continued prosperity. He also met Valentin

Soviet record album of Yevgeny Yevtushenko's BAM poem, 1977. (*Author's collection*)

Mokrovitsky, one of the chief engineers in the early days of BAM, an experienced railway constructor, who praised the involvement of the Komsomol in the project and told him: 'The Baikal–Amur Railway really does provide a wonderful opportunity for young people to express and prove themselves' (Ibid., p52). However, railway building could not be undertaken by enthusiasm alone and the amateur nature of much of his workforce must have been a frustration to Mokrovitsky and his professional colleagues.

The harsh and primitive conditions soon took their toll. Keenness among those who stayed for their contractual periods of work quickly ebbed when they witnessed the poor management of human and physical resources. Bullying and criminal activity were commonplace, as was sexist behaviour, sexual aggression and rape. Women soon found themselves undertaking traditional gendered roles, but when together could at least look after one another. Hard drinking was so extensive that safety at work was often compromised. Policing of the project was scant and behaviours tolerated or excused that would not have been the case in the cities and towns of the Soviet Union. Because the project was being managed through Komsomol organizations on the ground, corruption was widespread and resources were stolen and squandered. The main attraction holding workers to the project was the very high wages and productivity bonuses. Workers who stayed were also given vouchers for car purchase, but these ultimately proved worthless. With pressure to meet targets for track completion, absence of care for the environment became a serious problem that resulted in debate at various levels – a feature absent in the Virgin Lands campaign of twenty years earlier, but one never fully resolved. Some Komsomol environmentalists on the ground struggled hard to engender awareness of the delicate balance of the ecosystem in the taiga and received little support from outside. Lake Baikal was seriously polluted with oil, copper and zinc through the 'dirty' operations of industrial enterprises and operations associated with BAM. At the time some argued that it might never recover. There was little or no official concern about discharges from shipping, sewage and runoff of toxic fluids from the large-scale tunnelling operations. (Much of the information in this paragraph is taken from Ward, 2009, who provides the only study in English of BAM and its workforce.)

In post-Soviet times many of those who took part as young Komsomol members, but now living in a very different society, looked back fondly on the hardships they endured. In May 2009 the Moscow-based RT News programme spoke to one veteran, Alexander Gradusov:

> We lived in an igloo-like house in −45 degree Celsius weather. It was hard but inspiring. My wife and I were married there and our daughter was born there. Despite the difficult conditions we had everything we needed for life, like the helicopter that came to take us to hospital when our girl was due to be born. Afterwards, having lived so long without simple luxuries like a store we would come to a shop and fill carts with food. That went on for a whole year until once again we were used to having everything at our disposal.
>
> (*RT*, May 2009)

The achievements were remarkable: Tynda, a small encampment that would be the hub of the railway, grew in size to become the administrative capital of BAM, with a population of some 60,000. Construction took place outwards from Tynda and towards it from Lake Baikal and the west. Two other cities and over a hundred other settlements of various sizes grew up along the railway between 1975 and 1985, and over one million people were reported to have moved into the BAM Zone, as it was known. By 1985, nine tunnels with a total length of more than twenty miles, and 4,200 bridges and culverts had been built. Laid with special rails that would withstand the climatic extremes, almost all of BAM's length rests on permafrost, assuming it would always be there. In post-Soviet times, there has again been a significant investment in BAM. The latter includes continued electrification of the western end of the line from Tayshet to Taksimo, and the building of a nine-mile tunnel at Severomuysky, work on which commenced in 1975 but was not completed until 2003, negating the need for a thirty-three-mile steep section of line. Some sixty settlements, planned as population centres, sprung up along the line, but many have since suffered complete depletion due to unemployment once BAM was completed and the construction gangs moved on. The economic dreams of BAM's planners were never realized in Soviet times, and by the late 1980s, it quickly became an embarrassing and forgotten backwater, associated disparagingly with an ageing, vain and senile Brezhnev who died in 1982. However, in the last twenty years, it has undergone a transformation and could become the

Posed image of Kunerma station, 1985. According to Russian census data, the population had declined from 589 in 1989 to 59 by 2010. (навстречу времени. От Байкала до Амура)

important conduit and developmental hub its Soviet planners envisaged. BAM, though, is threatened by climate change if the permafrost that supports the line melts and it sinks into oblivion.

* * *

The purpose of including this chapter in the book is to illustrate the similarity between the two major Soviet state-sponsored prestige projects described and the space programme. All three had economic and strategic goals, but very importantly, also represented the idealistic aspirations of the USSR, and its attempts to bring on board its people. They were also designed to showcase the results to the rest of the world, reassure those in the satellite and friendly countries of Soviet progress on the road to communism, and demonstrate the concrete and lasting nature of Soviet achievements to capitalist Cold War competitors. All were flawed, but their ultimate failures in delivery were all the result of over-extension and under-investment at a time when the economy was under strain. The idealism behind them remains their positive testament and legacy, as do some of the achievements – whether collective or individual. The ultimate failure was that of the USSR and its attempt, over a seventy-four-year period, to build socialism that would lead to lasting communism. None of these prestige projects contributed to that outcome, although all suffered from the same contradictions and afflictions that had plagued the Soviet state since its early years, and particularly those of the Cold War era – the subject of the next and final chapter.

Yevgeny Yevtushenko, in one of the final verses in his epic poem about BAM, wrote:

> There are false paths and there are false prophets.
> Those who lie to us will soon be gone.
> It is not the road itself, but the journey,
> And where it takes us.
>
> (Pitzmann.ru)

'Waiting for the Komsomol Construction Team', Moldova, early 1980s (*I. Chibzee via Wikimedia Commons*)

Chapter 9

The End of the Soviet Dream

Imagine a country that flies into space, launches Sputniks, creates such a defence system, and it can't resolve the problem of women's pantyhose. There's no toothpaste, no soap powder, not the basic necessities of life. It was incredible and humiliating to work in such a government.
Mikhail Gorbachev (2001) discussing the planned commission to solve the shortage of women's pantyhose in the late Brezhnev years

This story concludes in 1991 when the USSR was officially declared non-existent – Russia and the republics that remained under Moscow's control were renamed the Russian Federation. The dream of the perfect society, based on the principles of communism, for which many had sacrificed so much – in civil wars, hardships at home, world wars and competition with enemies determined to eradicate it – had suddenly evaporated. To those of us who followed events as they unfolded in the late 1980s and early 1990s, it all seemed quite incredible. With the mother country's blessing, the satellite countries, and many of the Soviet republics, one by one went their own way. What was left imploded and the ruling party, in power since 1917, was declared illegal. Almost overnight the Soviet monolith disappeared, leaving behind crumbling monuments, a surplus of obsolete military hardware, and memories of social security and employment. In place of the old system and its values arose a new class of super-rich oligarchs and their political friends, along with hitherto unknown levels of poverty and joblessness. Ethnic conflict and civil war also followed in its wake. In compensation, citizens were offered Western-style political freedoms and choice over imported consumer goods that were out of reach for most. For most citizens, there was a weariness about the state of the USSR, but fear of change and the uncertainty this would bring with it.

The principal precursors of the decline of Soviet hegemony will be described briefly in this final chapter. The stagnation of the economy and decline in living standards, the destruction of Korean Airlines Flight 007, war in Afghanistan, the Chernobyl catastrophe, and growing dissent in Poland and other satellite countries. These did not occur in sequential or accumulative order, but are the primary factors associated, by historians, with the decline and fall of the USSR in its final years. Significantly, they all had a bearing on how Soviet citizens felt about their country and its political system and contributed to a general lowering of confidence and self-belief. Political dissent in the USSR was treated as a mental illness but grew anyway – Andropov's internal security agencies in the early 1980s reckoned there were 8.5 million potential plotters and regime opponents, but could find no evidence that they were coordinated (Lewin 2005). Eventually, their views reflected the opinions of enough citizens to ensure the end of the regime.

It's the Economy, Stupid: Stagnation and Decline in the USSR

Chapter 1 described the slow but steady increase in living standards in the USSR after the Second World War. While starting from a low base due to the devastation caused by the war, the country recovered and life did begin to get better for most citizens – especially in the cities. The rise in prestige, associated with the space successes, was accompanied by a belief that progress would continue and that the road to communism (coming by 1980 according to Khrushchev) would mean better rewards for its participants. This is not what happened. As noted in the previous chapter, food production remained problematic and shortages leading to queues in shops for basic items were a constant feature. This continued into the 1970s and 1980s – agricultural output declined due to inefficiencies and recurring climate-related poor harvests, forcing the spending of currency reserves on food imports to make up for the most chronic deficiencies. This is illustrated by estimates compiled by the CIA in Cold War secret briefings (and since released) for the US government. They suggest that a farmworker in the USSR in 1960 produced enough output to meet the needs of five persons; this had barely increased by 1970. In the US the equivalent figures were over thirty for 1960, and almost twice that for 1970. For 1983 the differences had multiplied: the USSR's farmworker could now supply enough for nine others, while in the US the figure had leapt to sixty-six (CIA 1985). Food shortages impacted on the nation's health. Despite reasonably good free health services whose scope and quality had risen steadily since the 1917 revolution, health generally, measured by such factors as infant mortality and life expectancy, declined after 1970 (Brainerd 2006). This happened despite increases per capita in income and some improved social benefits brought in by Brezhnev.

Industrial decline mirrored problems in agriculture. Khrushchev had declared, at the triumphant 1961 Party Congress, that the Soviet Union would have a higher standard of living than any capitalist country by 1981. Although the Soviet economy demonstrated faster growth than the USA's between the 1960s and the mid-1970s, it still lagged far behind, reaching 57 per cent of that of the US at its zenith, and falling steadily afterwards, so that by 1984 its comparison level was 51 per cent (Ibid.). The production of the consumer items that Soviet citizens so desired never achieved planning targets: cars, washing machines and televisions were rare when they had long since become household items to most workers in the West. The poor supply of goods lowered labour output as there was little incentive to work harder and earn more if there was nothing to buy in the shops. Alongside all this, and contributing to the malaise, was a flourishing second economy of privately produced (or clandestinely manufactured) and stolen goods – a more reliable source for things that were needed than official ones. Western goods were sought after by the young – especially fashion clothing like denim jeans, and records of popular Western music (Yurchak 2005). Those that found their way in through the Baltic ports of Estonia and Latvia (with easy access to Finland and Sweden), and Eastern European countries, had a profitable and ready market. The quality of manufacturing in the USSR was generally poor, finished goods were badly designed and unreliable – the best, like the cars based on Italian Fiat models, were produced under licence. The technology gap between East and West widened rapidly and when the Soviets were forced to increase international trade in the 1970s, it was only their energy products (oil and gas) that found a lucrative market outside the satellite countries.

Brezhnev died of old age and various health conditions in 1982. He was succeeded, in quick succession, by his peers and deputies Yuri Andropov and Konstantin Chernenko, both of whom died after brief periods in power. In March 1985, leadership was assumed by Mikhail

The End of the Soviet Dream 177

Nikitsky household's 'Zarya' TV and radio sets, Kharkov, Ukraine, 1960. (*Nikitsky family*)

Gorbachev. Gorbachev was a devout communist, but also a realist who realized that the country's problems required urgent attention and that reform was necessary if socialism was to survive – a process he described as *perestroika*. This would involve changes to the centrally planned economy and not its replacement by the market. Improvements would come about through better involvement of workers in production plans, more openness (*glasnost*), reduction in bureaucracy, accelerated production with improved technology, and an overall plan to bring the USSR up to the levels of the US. As time went on market reforms were introduced that began to transform people's outlooks and disprove those who wanted to ditch socialism and communism. Gorbachev differed from his predecessors in his rejection of the cult-like status associated with previous leaders, particularly Stalin and Brezhnev whom he criticized heavily. He took time to break down barriers with ordinary workers and took their complaints seriously. He also tried, without success, to combat the nation's increasing alcohol misuse through campaigns and price rises – which did not make him popular. Gorbachev was in favour of the USSR's peaceful space programme but was in no position to fund it effectively. Thus, its continuation under his rule was limited to the relatively modest programmes already underway.

Korean Airlines Flight 007

Gorbachev's attempts to end the Cold War were discussed in Chapter 3. One significant event that spurred this and reduced popular Soviet belief in the integrity of its armed forces was the shooting down of Korean Airlines Flight 007 on 1 September 1983. This Boeing 747 aircraft was flying from New York to Seoul, South Korea, with 269 passengers and crew. In the early evening, the Korean pilots made a navigational error, going off course into restricted Soviet airspace over the Kamchatka Peninsula and were subsequently shot down by Soviet fighter planes. At the time Cold War tensions were high and both sides already believed the other was behaving aggressively and that armed conflict might soon arise. The incident was immediately used by US President Ronald Reagan to ramp up the war of words and condemn the USSR for having deliberately downed a civilian airliner. The whole incident reflected badly on the Soviets who initially tried to deny any responsibility, despite clear evidence to the contrary. Under Andropov's leadership, which if anything was less conciliatory than Brezhnev's, there were to be no compromises about the USSR's right to guard its airspace and defend its territory.

It took many years for the truth to emerge about the incident since the Soviets had found the aircraft in the Sea of Japan and had recovered its flight recorder. After being crippled by a Soviet air-to-air missile, the plane had stayed airborne for five minutes before the pilots lost control. It spiralled into the ocean. At the time there was a dispute about the rescue missions and who should conduct them, but efforts were too late to save any of those on board. The Soviets were confused about the incident. They had tried to communicate with the airliner and had fired warning shots which had been ignored, and were not initially sure that they had shot it down at all. What also emerged at a later date was that the Americans had a spy plane in the area that was making deliberate incursions over Soviet territory and that its presence had been picked up by Soviet air-warning systems. Soviet Air Force fighters were alert to its presence and looking out for it. The attack was based on orders stating that if this was a military aircraft it should be shot down, but clear identification was necessary. Soviet communication records confirm this. In the dark, visual identification was impossible and the attack took place on the basis that the aircraft must have had aggressive intentions.

The disastrous outcome was therefore accidental. Both sides had some justification for believing what they did at the time, but with hindsight, and as the facts emerged over the following ten years, it reflected badly on both. When the Americans shot down an Iranian civilian airliner in 1988, killing everyone on board, they blamed problems with operating complex military technology – an excuse they had denied to the Soviets five years earlier.

The downing of Flight 007 served as a propaganda coup for the Americans and gave them the excuse to take a hard line with the Soviets when they met for a scheduled summit in Madrid a few days later. It gave Reagan all the support he needed at home to move his Pershing Cruise missiles into Europe and escalate the Cold War. As we saw in Chapter 3, this eventually pressured the Soviets under Gorbachev into disarmament talks that were largely conducted on American terms. The impact on the Soviet people was one of alarm and genuine fear of war. Criticism of the state's actions at that time was difficult but, when Gorbachev later went into talks with Reagan to end the Cold War, he enjoyed genuine popular support.

The War in Afghanistan

In the 1960s and 1970s, with the help of Soviet influence and investment, Afghanistan began to emerge from its traditional backwardness into the twentieth century. The king, who had ruled since 1933, had tentatively begun this process but was overthrown by his cousin, who established a republic in 1973. The rule of government, just as in the past, barely extended beyond the cities where factions vied for control among the new ruling clique. In April 1978 the pro-Soviet Communist Party, itself split with bitter internal division, seized power in a coup and began far-reaching reforms that were anathema to the Muslim religious figures who still enjoyed great influence over most of Afghanistan. Some, like the emancipation of women and rapid growth in educational opportunities for all, were progressive from a Western point of view, but deeply offensive to many traditionalists on whom they were forcibly imposed. Land reforms that divided rural communities were bitterly opposed because they put a sudden and an unnegotiated end to long-standing social relationships and hierarchies. Reform was backed up by repression, and opponents were jailed and murdered. Meanwhile, links with the USSR were formally consolidated and Soviet advisers appeared in military and civilian settings. In response, large numbers of men from the villages took to the hills to organize resistance in, as yet, uncoordinated groups with loyalties to local tribal leaders. In the cities and towns under government control, factions within the ruling party threatened any stability that might have emerged over time.

Ten years earlier, Brezhnev had justified his invasion of Czechoslovakia by declaring that once a country had become 'socialist' (as defined by the Soviet Union), other socialist countries had a duty to defend its gains if these became threatened by forces, internal as well as external: the Brezhnev Doctrine. He now used this to move to consolidate the gains of the 'Afghan Revolution'. In December 1979, 100,000 Soviet troops moved into the country, spearheaded by special forces who murdered the Communist leader Amin, who was blamed for the extreme policies that had caused dissent. In his place, the Soviets installed their own placeman, the more moderate communist Barbak Karmal. Just as in the nineteenth century when Tsarist Russia had interfered in Afghan affairs to forestall British influence, the Soviets had an eye on their southern neighbours that had been subdued and won over to socialism and communism in the 1920s. Their goal, as then, was to achieve an accommodation between socialism and Islam. The aim was for a short sharp intervention that would stabilize the regime and then a rapid troop withdrawal within six months. This was not to be.

The Soviet invasion, as characterized by the Americans and their allies, was condemned and became a Cold War proxy battleground that bogged Soviet troops down in a bitter conflict for the next nine years. Government control never extended beyond the major cities, leaving most of the country in the hands of competing guerrilla *mujahideen* groups, who, with the benefit of arms and supplies from neighbouring Pakistan (mostly paid for by the US and countries such as Britain), were able to conduct a war that could last indefinitely. Soviet troops were scattered across the country in helicopter-supplied enclaves to guard communication routes. This often meant long spells of duty by small groups of soldiers in remote areas where the threat of attack was constant. Back home there was little popular support for the intervention and many troops felt themselves forgotten, neglected, and, like their American counterparts in Vietnam

The last Soviet troops leave Afghanistan, 15 February 1989. (*A. Solomonov RIAN via Wikimedia Commons*)

ten years earlier, fighting an unjustifiable war where meaning could only be found in loyalty to immediate comrades.

The long war reached a climax in 1985 with Soviet offensives that tried to break the impasse but which only served to rally resistance.

Atrocities occurred from both sides and civilian deaths were estimated to be well over a million (Sliwinski 1989). The Soviets, like the Americans in Vietnam, used unacceptable means to deny the mujahideen bases and support, and destruction of rural infrastructure and cultivatable land was widespread. Soviet military deaths amounted to 14,453 with a further 53,753 wounded, 10,751 of whom were left permanently disabled. Among themselves, Soviet troops in Afghanistan were known as Zinky Boys after the zinc coffins in which the dead were returned to the USSR (Alexievich 2017).

Changes in the USSR, mounting opposition, unaffordable costs, and military stalemate, led to a complete, internationally negotiated withdrawal of Soviet troops by February 1989. The Soviet mission had failed and its legacy in creating several generations of experienced and well-armed mujahideen fighters, was lasting – as was the devastation. Ironically the Stinger ground-to-air missiles, supplied in vast numbers by the Americans to the guerrillas to shoot down Soviet helicopters, were turned against US and Allied forces when they moved in to unsuccessfully impose their version of stability in later years.

Chernobyl

At the peak of its success and for many years afterwards, the USSR's space programme was a testament to the high levels Soviet science and technology had reached. Scientific progress was also demonstrated through the development of nuclear power: the USSR was the first country to build a nuclear power station, at Obninsk, seventy miles southwest of Moscow, which opened in June 1954. This was projected to Soviet citizens, and across the world, as proof that atomic science had non-military use of benefit to all mankind – the communist ideal of peace and progress. Association with the production of nuclear weapons was omitted from the propaganda; instead, international exhibitions showcased the nuclear-powered icebreaker *Lenin*, power generation and the spread of electrification in the USSR. There was certainly a popular belief, despite the hardships of life that many experienced, that Soviet technology, because it was not driven by market forces, was superior even if its benefits were out of reach for most people. The notion of 'Atoms for Peace' was coined by US President Eisenhower in a speech to the

United Nations in 1953, launching an educational and nuclear technology assistance programme. The Americans, of course, just as much as the Soviets, needed a cover for their continued development of nuclear weapons, which was the subject of intense Cold War competition.

Soviet dreams about cheap people's energy were to be shattered by the events of 28 April 1986, when the No. 4 reactor at Chernobyl in the north of Ukraine suffered disaster with devastating consequences – the worst nuclear accident in history. By April 1986 the Soviet Union had forty-five nuclear reactors spread over seventeen sites. Two early ones had already stopped operating and many more were in the process of commissioning or in the planning stages. By contrast, the US in April 1986 had 113 reactors over seventy-eight sites – a significantly greater investment (IAEA 2006). Safety had always been a concern, and serious incidents which resulted in radioactive fallout at Windscale in the United Kingdom in 1957 and Three Mile Island, US, in 1979, underlined the fallibility of nuclear power as an energy source.

The Chernobyl disaster was a consequence of human error when a safety test went wrong, and reactor design flaws that escalated an already-deteriorating situation once meltdown was underway. The

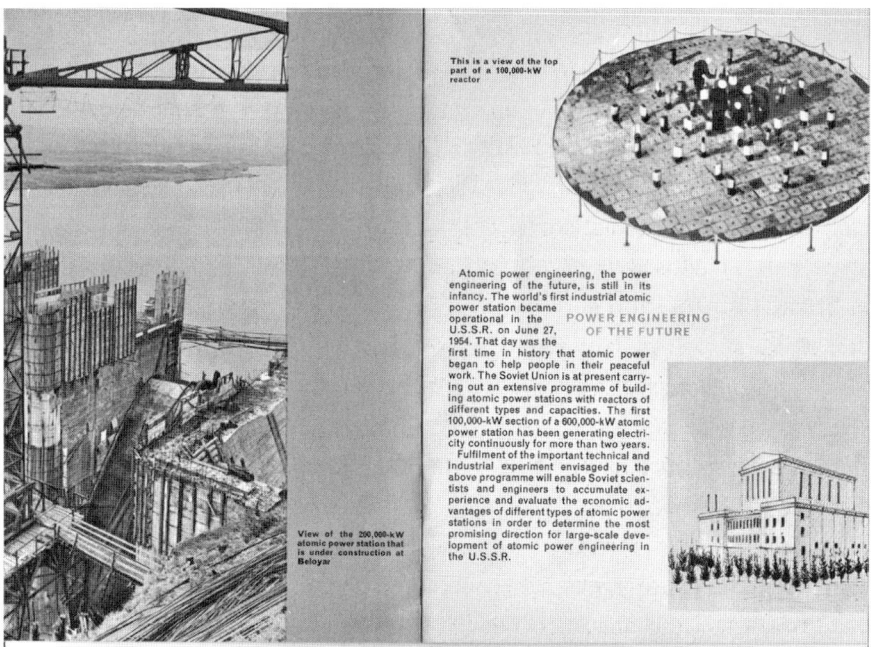

Pages from *Peaceful Atom*, a Soviet London Exhibition booklet, 1961. (*Author's collection*)

nuclear chain reaction got out of control quickly, causing damage that allowed huge amounts of radioactive fallout to escape and spread west and north on the prevailing wind into neighbouring countries and across Europe as far as Britain and Ireland. In typical fashion, the Soviet authorities initially denied the extent of the accident and downplayed its consequences. When this response became unsustainable, as the radioactive cloud was detected falling over Sweden, they began to tell the truth and face up to the consequences. After an initial delay, the rapid and permanent evacuation of the nearby city of Pripyat's 50,000 population was enacted. Up to 600,000 Soviet firefighters, police officers, soldiers and volunteers were deployed to help fight the fire, and clean up the area. Many died knowing the risks (Chernobyl Forum 2003–2005). They were collectively known as Liquidators and were officially revered as heroes. The immediate operation involved disposing of radioactive debris from the explosion and pouring concrete by helicopter into the remains of the reactor to seal it from further leakage into the air. Attention was also paid to nearby watercourses and any other means by which contamination might spread. This was a vast and hugely expensive operation that continues to this day – work is still underway to permanently seal off reactor No. 4. Expenditure worldwide over the two decades following the disaster was estimated to be in the region of hundreds of billions of dollars. In 1991 Belarus (as the Byelorussian SSR had become known), just five years after the disaster, spent a fifth of its annual budget on Chernobyl-related activities (Chernobyl Forum 2003–2005).

The cloud of radioactive dust affected 38,610 square miles of land, a large proportion of which was in the neighbouring Byelorussian SSR (Marples 1996). However, some perspective can be gained by the fact that although there was said to be 400 times more radioactive material released than the in the bombing of Hiroshima and Nagasaki combined, it still only represented between a hundredth and a thousandth of that released during the nuclear weapons tests at the height of the Cold War (IAEA 1997). The human cost of Chernobyl is difficult to calculate: thirty men died in the initial explosion and from exposure to radiation in the immediate aftermath of the accident, and up to 6,000 of the Liquidators may have died in the following decade as a consequence of their work (Marples 1996). Cancer-related deaths attributed to Chernobyl

The End of the Soviet Dream 185

Helicopter spraying decontamination liquid, Chernobyl, 13 June 1986. (*IAEA via Wikimedia Commons*)

are contested but the environmental organization Greenpeace estimate that these have reached 100,000 (Greenpeace 2019), including cancer-suffering children unborn in 1986.

For the USSR after 1986, an overwhelming cost of Chernobyl was in the loss of confidence in the leadership and officialdom generally. Denials and untruths emanated from official sources from the start, and then individuals associated with Chernobyl's operations were held responsible for mostly systemic failures. Health information was misleading and impacts minimized to the extent that foreign aid in the aftermath was refused out of nothing more than Soviet pride. Lower-level officials initially presented optimistic reports lest they be blamed for mistakes that had contributed to events – leading to confusion among the leadership. All this led to mass distrust, especially in the areas most affected, about all official information – a legacy that continued long afterwards (Chernobyl Forum 2003–2005). The belief in Soviet science and technology was seriously dented and the arguments of Western environmentalists about the dangers of nuclear power took root among many in the younger generation. Cynicism grew and became rampant:

At that period, my idea of a nuclear power station was quite idyllic. At school and in college, we were taught that these were 'fairy tale factories making energy out of nothing', in which people in white coats sat and pushed buttons. Chernobyl exploded in minds which were completely unprepared, which had complete faith in technology … That was made worse by the total lack of information. Rumour was rife: someone had read in a newspaper, someone had heard somewhere, someone had been told … Even those trying to extinguish the reactor were, it turned out, dependent on rumours.

(Zoya Danilovna Bruk, nature conservation inspector, quoted in Alexievich 2016b, p208)

Gorbachev, in post for just over a year, was well aware of all this and took public opinion into account in his calculations about the way forward. His policy of glasnost, however, was revealed for its shallowness as the disaster unfolded, but the experience certainly accelerated the reform programme. It was three weeks before Gorbachev addressed the nation on TV about the disaster. He later said that he thought that Chernobyl was the most significant single contributor to the downfall of the Soviet Union (*History is Now* magazine, 2017).

Unrest in the Satellite Countries

The imposition of Soviet-dominated socialist regimes in the countries of Eastern Europe in the aftermath of the Second World War was never achieved with the full consent of their people (with the arguable exception of Czechoslovakia). The fact that once in place, these regimes were both immovable and unreformable, caused a simmering discontent that surfaced from time to time. Typically, unrest was spurred by sudden falls in living standards. Disorder took place on this basis in East Germany in 1953, Poland in 1956 and on numerous occasions in the 1970s and 1980s. Additionally, there were instances of unrest provoked by the absence of democracy and basic freedoms. This led to serious popular uprisings in Hungary in 1956 and Czechoslovakia in 1968 when the ruling party moved to reform itself, ending on both occasions with military intervention by the USSR. Although the Soviet-leaning United Workers' Party in Poland never bent to popular pressure until the late

1980s, unrest in Poland united workers organized in the independent trade union Solidarity, with intellectuals who wanted political freedoms. Poland became the major centre and focus of satellite-country unrest from the 1970s onwards.

The military interventions by the Red Army in 1956 and 1968 caused a major upset in the international communist movement. On both occasions, loyalty to the Soviet Union was dented with mass resignations from Western communist parties (Hungary, 1956), and shifts in position that moved them away from automatic support for Soviet policies (Czechoslovakia, 1968). The Brezhnev Doctrine was used to justify the direct military intervention in Afghan affairs, but in the same era, never to quell unrest in Poland. The reasons for this are quite clear: at no time did the rulers of Poland seek to reform themselves. Instead and almost to the end, they sought to fight and contain the forces of change through their own means – including the use of police violence, mass arrests, detention, and other forms of oppression. By the time the pressure for change had become so popular that only decisive military intervention from the USSR might have changed the course of events, it was too late, as Gorbachev had moved the Soviet Union a long way from such interventions. With

Captured Soviet tank, Budapest, Hungary, 1956. (*Unknown photographer via Wikimedia Commons*)

Scene at the gates during the strike at the Lenin Shipyard, Gdansk, August 1980.
(*L. Szmaglik ESC via Wikimedia Commons*)

no one to save them, the leadership of the United Workers' Party meekly gave way to democracy. Free elections took place in June 1989, bringing to an end forty-four years of pro-Soviet dictatorship. The country quickly turned westward in its outlook.

The movement for change in Poland had ripples throughout the Soviet Bloc as it grew in strength through the 1980s. The other nations that had never accepted Soviet rule very happily were the Baltic republics of Estonia, Latvia, and Lithuania. Soon protest movements based purely around nationalist demands for freedom took off in popular form. In one potent demonstration of support in August 1989, two million people formed a 600-kilometre human chain from Vilnius in Lithuania, through Riga in Latvia and up to Tallinn in Estonia. In November 1989 these three countries were accorded autonomy to take effect from 1 January 1990. They too turned to the West for inspiration and links.

The most symbolic evidence of the cracks in the USSR's dominance over Eastern Europe took place in November 1989 when the Berlin Wall was dismantled, a precursor to the reunification of Germany (or rather the absorption of the German Democratic Republic into the Federal Republic of Germany). Other pro-Soviet regimes and Stalinist dictatorships across Europe fell one after the other – not a single one remained loyal to the USSR. Within the USSR, the secession of the Baltic republics was followed by that of Ukraine, Byelorussia and those in the south of the USSR: Georgia, Azerbaijan, Armenia, Kazakhstan, Tajikistan, Uzbekistan and Moldavia. While this process was complete by the beginning of 1993, civil wars and ethnic conflicts broke out across the eastern areas of the region, and in the former Yugoslavia. Old enmities were revived in the name of religious difference and territorial dispute, thousands dying and tensions continuing to simmer in many areas well into the new century.

The End of the USSR … As Seen from Space

This book's central theme has been the USSR's space programme and the successes of the 1950s and 1960s. The space programme was not immune to the changes taking place in the country. In August 1991, a coup by old-guard establishment members opposed to Gorbachev's reforms tried unsuccessfully to dislodge him. This failed when most of the military remained loyal and stood by the thousands of ordinary people who took to the streets in Moscow to defend their leaders and the programme that was underway to renew the Soviet Union. Gorbachev was on holiday and a leadership role was taken by Boris Yeltsin who emerged a hero. Although Gorbachev was severely weakened and resigned a few days later as general secretary of the CPSU, he remained as president. The victors, led by Yeltsin, pushed through further reforms including the banning of the CPSU, signalling the end of Soviet power. The finale came on 26 December 1991 when the Supreme Soviet voted to dissolve itself. The Soviet flag was lowered for the last time, Gorbachev resigned and handed over the presidency of the new Russian Federation to Boris Yeltsin.

The fortunes of the USSR at this time were reflected in space. NPO Energia, the central agency, as OKB-1 had become in 1974, had by this time entered into partnerships with anyone willing to pay, and several astronauts from other nations (including the US) had been taken into space from Baikonur and involved in the activities of the Mir space station. On 19 May 1991, Soyuz TM-12 was launched to take a group to Mir for an extended stay. On board were Anatoli Artsebarsky as commander, Sergei Krikalev as flight engineer, and Helen Sharman from Great Britain (the first British astronaut). Sharman returned to Earth on Soyuz TM-11

Soviet tank guarding the parliament building during the failed coup, Moscow, 19 August 1991. (*I. Simochkin via Wikimedia Commons*)

with the crew already aboard Mir, a week later. Krikalev and Artsebarsky remained on the space station to conduct further experiments. The next mission on board Soyuz TM-13 arrived on 2 October – two missions had been combined into one at this stage for cost reasons: aboard were an experienced Soviet cosmonaut, Alexander Volkov, and a Kazakh, Toktar Aubakirov, whose presence seems to have been symbolic, a means of ensuring that the soon-to-be independent Kazakhstan continued to permit Soviet use of the Baikonur cosmodrome. The third person on the mission was Austrian Franz Viehböck, whose government had paid NPO Energia $7 million for the privilege. According to at least one report, the presence of the Austrian was a substitute for the cosmonaut who might have relieved Krikalev – income at this stage was imperative for NPO Energia (*Discover magazine*, December 2016). Artsebarsky returned to Earth with Viehböck on Soyuz TM-12 on 10 October leaving Volkov and Krikalev on Mir. Krikalev was in regular communication throughout this period with a Russian-speaking amateur Australian radio enthusiast, US-born Margaret Iaquinto. She kept him and Volkov informed about the unfolding events in the Soviet Union. Krikalev's stay in space eventually stretched to a record-breaking 336 days – twice as long as planned when

Gagarin Terrace, Kilwinning, Scotland, October 2020. The residents of this house did not know the origin of their street name. (*A. Anderson*)

he entered Mir in May 1991. He and Volkov eventually returned to Earth on Soyuz TM-13 on 25 March 1992, emerging in their hammer-and-sickle-badged spacesuits to a landing site in what had become a foreign country (Kazakhstan). They were met by a recovery team under the new red, white and blue flag of the Russian Federation.

By the end of the decade, the economy of the Russian Federation was in tatters: GDP had fallen to almost half its 1980s value. Yeltsin's policy to maintain the economy consisted of selling its most viable assets (in particular energy production) for rock bottom prices to his friends – the new breed of oligarchs, many of whom had been functionaries in the Soviet regime and understood the potential market value of its state-owned corporations (Ghodsee 2017). In 1994 NPO Energia was partially privatized to become SP Korolev RKK Energia. Quite what Korolev would have thought will never be known, but it is unlikely he would have been impressed with this use of his name. He had been a loyal Soviet citizen, and despite his indifference to politics, he might have taken issue with the impoverishment of most Soviet citizens, and the subservience of its space programme to an international market. Gagarin, who was central to Soviet success, was ranked sixth in a 2010 American survey of all-time space heroes – this, jointly with the fictional Captain Kirk of the Starship Enterprise (Doran and Bizony 2011) – surely a true sign of the marginalization by then of the significance and achievements of the Soviet space programme.

In 1918 Lenin had told the Bolshevik Party congress:

> Regarded from the world-historical point of view, there would doubtlessly be no hope of the ultimate victory of our revolution if it were to remain alone, if there were no revolutionary movements in other countries. I repeat, our salvation from all these difficulties is an all-Europe revolution. At all events, under all conceivable circumstances, if the German revolution does not come, we are doomed. ... But if it does not turn out as we desire, if [revolution] does not achieve victory tomorrow – what then? Then the masses will say to you, you acted like gamblers – you staked everything on a fortunate turn of events that did not take place, you proved to be unequal to the situation that actually arose instead of the world revolution, which will inevitably come, but which has not yet reached maturity.
>
> (Lenin, 1918)

Of course, Lenin could not have foreseen the turn of events some seventy years later after so much water, much of it tainted and bloody, had passed under the Bolshevik bridge to communism, but his words proved remarkably prophetic. The world failed to follow Russia's lead in 1917 – the rest is history. The achievements of the Soviet Union were nonetheless remarkable and these include the space successes of the 1950s and 1960s, and ironically, the USSR's contribution to peace in the world. The loss of the balance of power maintained throughout the Cold War, despite the proxy wars and wasteful arms spending, was soon followed by global instability and conflicts that continue. Like so much in history, it could all have been very different.

It is difficult to imagine the confusion, heartbreak and desolation that the sudden break-up of the USSR caused to so many who had known nothing else and had believed in the state and its ideals, even if they might scorn them in private. What had gone before was 'normal' life and they had just got on with it, never expecting that the uncertainty that 1991 brought could happen in their lifetimes. Some of the mixed feelings, elation and despair, have been captured by Svetlana Alexievich in her recording of eyewitness and personal accounts (2016b). That year Party loyalists dumped their Party cards and Komsomol certificates outside official offices at night, denying everything they had stood for and believed in. Within a few years, many were bitterly disappointed with what change had brought them and longed for the certainties and securities of the Soviet times. Their world was gone forever and now the people they most feared were not American but those at home who cared little of how they were faring and were happy to exploit and steal from them, legally or otherwise.

Now the people of the Soviet Union were thrown into the melting pot of a market system where everyone had to take care of themselves and where the collective ideals that promised a better world were gone forever. Gagarin's words from an era that was being rapidly forgotten seem particularly appropriate to close.

> When I orbited the Earth in a spaceship, I saw for the first time how beautiful our planet is. Mankind, let us preserve and increase this beauty, and not destroy it!

18 МАРТА 1965 ГОДА ЧЕЛОВЕК ВЫШЕЛ В ПРОСТОРЫ ВСЕЛЕННОЙ.

1965 Soviet postcard of Leonov's first walk in space. (*Author's collection*)

Appendix

Leading Characters

Valentin Bondarenko (1937–1961) Soviet air force fighter pilot from Kharkov, Ukraine. One of the first cosmonaut trainees. His death in a fire during training in March 1961 made him the first fatal casualty of the USSR's space programme. His death was kept hidden for almost twenty years.

Leonid Brezhnev (1906–1982) Communist Party member from 1923, and Red Army officer who rose to the rank of major general in the Second World War. He became a member of the Central Committee of the CPSU in 1952 and the Politburo (the leading inner circle) in 1957. He was responsible for implementing the Virgin Lands Campaign in Kazakhstan. He was instrumental in ousting Khrushchev in 1964, after which he became first secretary (the most powerful position in the USSR), a position held until his death in 1982, by which time his name was synonymous with self-aggrandisement and the stagnation of the economy, but at the same time, détente and international stability. Brezhnev is associated with the BAM railway project, which he hoped would be his lasting legacy.

Wernher von Braun (1912–1977) Born in a German town near what later became Poznan in Poland, von Braun was interested in astronomy and rocketry from an early age. After achieving his doctorate in 1934 from university in Berlin, he pursued rocket design and propulsion engineering in (by now) Nazi Germany as a civilian within the military establishment. Von Braun joined the Nazi Party in 1937 and the SS (the principal Nazi paramilitary force) in 1940. By 1943, having been elevated to professorial status by Hitler, he was heading the team that developed V-1 and V-2 missiles launched against Britain and other Western European targets in the closing stages of the war. These were built in slave labour camps where thousands died. Captured by the Americans and taken to the US, von Braun became a US citizen and eventually head of the NASA space programme, his embarrassing Nazi past downplayed and glossed over as his technical expertise was exploited to the full.

Dwight D. Eisenhower (1890–1969) US career army officer who rose to become a popular five-star general and Allied supreme commander of the Allied Expeditionary Force in Europe. He entered politics as a moderate Republican, winning the presidency by landslide victories in 1952 and 1956. One of his main political aims was to reduce the spread of communism and his defence and aerospace policies centred on this. After Sputnik's success in 1957, he established NASA to escalate the US space programme – the start of the space race.

Valentina Gagarin (1935–2020) Born in Orenburg, Soviet Union, Valentina was the youngest of six children; two of her brothers were killed in action in the Second World War. After graduation from the Orenburg medical school, she worked as a laboratory assistant at the air force base medical centre. It was at a dance there that she met Yuri Gagarin who was in the early stages of pilot training. The couple completed their respective studies before marrying in 1957. Valentina was a modest, shy and very loyal individual who tried to stay in the background when her husband became famous in 1961. She did have to accompany him on some trips but preferred life at home raising their two daughters. After Yuri died in 1968, and a widow at 32, she stayed on in Star City, working in the medical centre and later writing a book about her husband. Valentina refused most interview requests and avoided publicity. If her husband was a hero, so too was Valentina.

Yuri Gagarin (1934–1968) Almost the opposite of his wife, Gagarin was a vivacious and determined character whose charm and tact made him the perfect candidate to be the first man in space. Born in a village near Gzhatsk, Gagarin endured severe hardship as a child in the Second World War, his home area being under occupation by German troops. On leaving school he became a foundryman but eventually followed his dream and joined the air force to train as a pilot, graduating in 1957. In 1959 he was selected for the Cosmonaut training programme and, with the support of his colleagues, was chosen to be the first man in space for the projected flight in April 1961. Fame followed this, which eventually took its toll on his health. Gagarin was still hoping to return to space when he was killed in a training flight in 1968.

Valentin Glushko (1908–1989) From Odessa, Ukraine, Glushko was another whose interest in rocketry and space started in childhood. After school, he worked first as a sheet metal worker before entering Leningrad University where he failed, bored with academic study, and left without a degree. Through the 1930s he worked in GIRD before arrest and imprisonment in 1938 as a victim of the purges. As a prisoner until 1944 he worked on rocket propulsion research, heading a team that included Korolev, and in 1945 was sent to Germany to gather information about the Nazi rocket programme. During the early years of the Soviet space programme, Glushko was responsible for engines that were used by Korolev. In 1974 he was placed in charge of the newly formed NPO Energia space agency, a post he held until his death.

Mikhail Gorbachev (1931–) From a poor peasant family in the Stavropol region of south-west Russia, Gorbachev started working life as a combine harvester driver on a collective farm, before studying at Moscow University, where he met his wife Raisa who was also a student, and who became a strong influence on his political outlook and direction. After graduation, he returned to Stavropol where he worked his way up through the Komsomol and Communist Party machines before going back to Moscow as a Central Committee member in 1978, and to the Politburo the following year. Gorbachev was the first secretary of the CPSU from 1985 until 1991, ushering in reforms that led directly to the downfall of the USSR. In 1991, he founded the International Foundation for Socio-Economic Studies (the Gorbachev Foundation) which continues to analyse Russian and world affairs.

Nikolai Kamanin (1908–1982) A pre-war flying hero and Second World War pilot and air commander, Kamanin was a career air force officer from Melenki, 200 miles east of Moscow. His son Arkady became a wartime pilot at the age of 14 but died of meningitis in 1947 when only 18. In 1960, Kamanin, now a general, was appointed to head the cosmonaut training programme. As such he was responsible for selecting and managing the cosmonauts, including intervening when necessary in their personal lives. He pushed them hard but was sympathetic to their problems. Kamanin retired in 1971, handing over his role to former cosmonaut Vladimir Shatalov. His diaries, published between 1995 and 1997, offer a fascinating insider insight into the Soviet space programme.

John F. Kennedy (1917–1963) From a powerful and wealthy Irish-American family, Kennedy was a Second World War hero and entered politics in 1947 as a congressman in the house of representatives. He narrowly won the presidency as a Democrat against Richard Nixon (Eisenhower's vice-president) in 1960, and continued his Republican predecessor's popular stand against communism, which included continued investment in the space programme after Gagarin's success in 1961, confronting the USSR in Cuba (narrowly avoiding nuclear war), and stepping up US involvement in Vietnam. Kennedy was assassinated in November 1963.

Sergei Korolev (1907–1966) From Zhitomir, Ukraine, Korolev is generally credited with the successes of the Soviet space programme during his lifetime, and his absence blamed for some of its failures afterwards. Fascinated by flying from an early age, he grew up developing an interest in rocketry and the possibility of space travel. After graduation from a technical university in Moscow, he worked in the aircraft design bureau and then at GIRD as head in 1932. He was arrested in 1938 and suffered badly in detention in the Kolyma goldmining camps before transfer to an aircraft design bureau under Tupolev in 1940. Korolev was not released until 1944 and from then was involved in OKB-1, retrieving German rocket designs for the ballistic missile programme. With his long-standing interest in space travel, he succeeded in gathering political support for the Soviet space programme which he led, as the publicly anonymous chief designer, until he died in 1966.

Nikita Khrushchev (1894–1971) From the poverty-stricken village of Kalinovka in western Russia, Khrushchev was the son of poor peasants, his father a sympathizer of the revolutionary movement and later a trade union organizer. After only four years of schooling, he started work as a herd boy and later, in a succession of manual labouring jobs, he became a skilled metalworker which gained him exemption from military service in the First World War. After the February 1917 revolution, Khrushchev was elected as chair of the workers' soviet (council) in Rutchenkovo. In 1918 he joined the Bolshevik Party and became a political commissar during the Civil War, the start of a lifelong career serving the CPSU. In the 1930s, as a rising apparatchik in Stalin's USSR, he was responsible for implementing purges, and during the war became a ruthless leading Party commissar in various important locations as the war progressed, including Stalingrad during the great battle. After Stalin's death, he was quick to manoeuvre himself into position as the new leader, forgetting his past as one of Stalin's most loyal servants and denouncing his crimes to steer the USSR on a new and less oppressive course. Khrushchev supported the space programme and ensured it received the investment required to beat the Americans in peaceful competition as the Cold War progressed. His policies were not popular with many in the leadership and he was unceremoniously removed in 1964 and almost written out of the USSR's history until the 1980s.

Alexei Leonov (1934–2019) Leonov was born in Litsvayanka, Western Siberia, where his grandfather had been exiled as a supporter of the 1905 revolution. He was one of nine children; his father, a mining electrician, was arrested in the purges the 1930s and declared an 'enemy of the people' before being released and rehabilitated. In 1948 the family relocated to Kaliningrad on the border with Poland in the far west of the USSR, an area which had been forcibly depopulated of its German residents at the end of the war. Leonov wanted to study art after finishing school but went for pilot training as the family could not afford the tuition fees. Leonov graduated as an air force fighter pilot in 1957 and was chosen in 1959 to be among the first cosmonaut trainees. In 1968, on his first space flight, he became the first person to walk in space outside his capsule. Leonov was selected for several other flights, all of which were cancelled (he would have been on the first Soviet moon mission had the entire programme not been axed). His second trip into space did not take place until 1975 – a joint American and Soviet mission. His art output was extensive and included numerous space-related works.

Ronald Reagan (1911–2004) From a poor northern Illinois family, Reagan worked as a radio sports commentator and then as a Hollywood B-movie star in the 1940s and 1950s. He became involved in American politics through his anti-communist activities as president of the Screen Actors Guild, becoming Republican governor of California in 1966. As a right-winger in his politics, Reagan preferred the simplistic good guys versus bad guys and straight-shooting approach of the cowboy characters he had once played to anything more complex. In 1979, he was elected Republican president of the US and immediately began to move against the détente that had existed for some years with the USSR, dangerously escalating the arms race.

Alan Shepard (1923–1998) From New Hampshire, a Second World War veteran, Shepard graduated as a US Navy pilot in 1946, going on to become a test pilot in 1950. He was selected for space training as one of the initial group of American astronauts in 1959 and then became the second man, and first American, in space in May 1961. He was grounded in 1963 due to an ear defect (Ménière's disease) but reinstated after this was corrected surgically in 1969. In 1971, he commanded the Apollo 14 mission to the moon and walked on the surface, playing the first extra-terrestrial and weightless game of golf when he fired off two shots.

Valentina Tereshkova (1937–) A textile worker from Yaroslavl whose ambition was to become a train driver, Tereshkova was a parachutist and skydiver through her local Komsomol organization in which she played a leading role. Her father had been killed in the Winter War with Finland in 1940 and her mother had raised the family alone. Selected for the small group of women cosmonaut trainees through the Komsomol, Tereshkova was one of the least academic and technically proficient of the group, but as a good parachutist and representative of the ideals of socialism, was considered the right candidate to be the first woman in space in a hurried programme to beat the Americans. Her mission in November 1963 was successful and, like Gagarin, she rose quickly from obscurity to worldwide fame. Tereshkova too had charm and good diplomatic skills that were deployed in her ambassadorial role for Soviet women – she made more trips abroad than any other cosmonaut. For a while, she was married to fellow cosmonaut Andrian Nikolayev. In the post-Soviet era Tereshkova entered politics and at the time of writing is an elected representative for Yaroslavl of the ruling United Russia Party.

Gherman Titov (1935–2000) Titov was born in a remote village in the Altai Krai region near the border with Kazakhstan where his father became a school teacher. As the second Soviet in space (and third human), Titov's mission was the most complex to date, involving multiple orbits, manual control of his craft and tasks such as photographing the Earth, proving that humans could spend extended periods outside the Earth's atmosphere. Titov remained involved in the space programme after his flight but never went back into space. He too was a popular ambassador at home and abroad, and first suggested that the date of Gagarin's flight should become Cosmonautics Day. In his later years, he entered politics in the post-Soviet Russian Federation.

Konstantin Tsiolkovsky (1857–1935) Although revered as the founding father of Soviet rocketry and space science due to his pioneering experiments and books that contained his ideas about possibilities for the future, this school teacher from Kaluga was never officially supported with his work and never benefited from being able to work in seats of learning. His scientific work was conducted in his spare time, motivated by enthusiasm alone. Tsiolkovsky's life was certainly remarkable – he was mainly self-taught as a child as the local school could not deal with his deafness. During his life he wrote prolifically on space science which influenced the generation of Soviet designers and engineers who put together the first programmes. At the very end of his life, he became officially recognized and lauded as his visions chimed with the USSR's promotion of possibilities for the communist future. However, his belief in the pseudo-science of eugenics was not popular in the Soviet Union (as it was in Nazi Germany) and he was treated ambiguously during his lifetime but celebrated as a pioneering hero in the years of Soviet space success.

Bibliography

Аджубей, А. Горюнов, Д.П., Ильичёв, Л.Ф., Сатюков, П.А., Сиволобов, М.А., Скуридин, Г.А. (eds.) (1961) Утро Космической Эры (*Morning of the Space Age*), Москва, Госполитиздат
Alexievich, S. (2016a) *Chernobyl Prayer* London, Penguin
Alexievich, S. (2016b) *Second-Hand Time* London, Fitzcarraldo Editions
Alexievich, S. (2017) *Boys in Zinc* London, Penguin
Andrews, J. & Siddiqi, A. (2011) *Into the Cosmos: Space Exploration and Soviet Culture* Pittsburgh, University of Pittsburgh Press
Applebaum, A. (2003) *Gulag: A History* London, Penguin
Applebaum, A. (2017) *Red Famine: Stalin's War on the Ukraine* London, Allen Lane
Beevor, A. (2002) *Berlin, the Downfall 1945* London, Viking
Belov, F. (1956) *The History of a Soviet Collective Farm* London, Routledge & Kegan Paul
Brainerd, E. (2006) *Reassessing the Standard of Living in the Soviet Union: An Analysis Using Archival and Anthropometric Data* web.williams.edu/Economics/faculty/brainerd-ussr.pdf
Brezhnev, L. (1978) *The Virgin Lands* Moscow, Progress Publishers
Brzezinski, M. (2007) *Red Moon Rising: Sputnik and the Rivalries that Ignited the Space Age* London, Bloomsbury
Burchett, W. (1980) *At the Barricades: The Memoirs of a Rebel Journalist* London, Quartet
Burchett, W. & Purdy, A. (1961) *Cosmonaut Yuri Gagarin: First Man in Space* London, Panther
Burchett, W. & Purdy, A. (1962) *Gherman Titov's Flight into Space* London, Panther
Caidin, M. (1959) *X-15: Man's Daring Flight into Space* New York, Rutledge
Caidin, M. (1961) *Man into Space* New York, Pyramid
Chamberlain, W. (1946) *Blueprint for World Conquest: The Official Communist Plan* Washington, Human Events
Chernobyl Forum (2003–2005) *Chernobyl's Legacy: Health, Environmental and Socio-economic Impacts* www.iaea.org/sites/default/files/chernobyl.pdf
Chertok, B. (2006) *Rockets and People Vol. 2 – Creating a Rocket Industry* Washington, NASA
Chertok, B. (2009) *Rockets and People Vol. 3 – Hot Days of the Cold War* Washington, NASA
CIA (August 1964) *Comparison of US and Estimated Soviet Spending for Space Programmes* at www.cia.gov/library/readingroom/docs/DOC_0000316255.pdf
CIA (October 1985) *A Comparison of the US and Soviet Economies* at www.cia.gov/library/readingroom/docs/DOC_0000497165.pdf
Clare, J. *One Hundred Russian Jokes* at www.johndclare.net/Russ12_Jokes.htm
CPSU (1961) *The Road to Communism: Documents of the 22nd Congress of the Communist Party of the Soviet Union* Moscow, Foreign Languages Publishing House
CPSU (1961) *Programme of the Communist Party of the Soviet Union* Moscow, Foreign Languages Publishing House
Данилкин, л. (2011) Юрий Гагарин (*Yuri Gagarin*) Москва, молодая гвардия
Dick, S. (ed.) (2008) *Remembering the Space Age* Washington, NASA
Discover magazine (20 December 2016) *The Last Soviet Citizen* at www.discovermagazine.com/the-sciences/the-last-soviet-citizen

Doran, J. & Bizony, P. (2011) *Starman: The Truth Behind the Legend of Yuri Gagarin* London, Bloomsbury

Фотоальбом 'Я чайка' (1966) Ярославль, Верхне Волжское

Field, M. (1988) 'Soviet Society and Communist Party Controls' in *Understanding Soviet Society* London, Allen & Unwin

Гагарин, Ю, Николаев, А, Попович, В, Быковский, В, Терешкова, В, (1964) 'Космонавмы рассказываюм Издательство' (*Cosmonauts tell their stories*) 'Детская литература' Москва

Garrison, J. & Pyare, S. (1983) *The Russian Threat: Its Myths and Realities* London, Gateway

Gavin, J. (1959) *War and Peace in the Space Age* London, Hutchinson

Ghodsee, K. (2017) *Red Hangover: Legacies of Twentieth-Century Communism* Durham NC, Duke University Press

Ghodsee, K. (2018) *Why Women Have Better Sex Under Socialism* London, The Bodley Head

Голованов, я. (Golovanov, Y.) (1994) Королев – факты и мифы (*Korolev – facts and myths*) Москва, Наука

Golubev, G. & Gronin, N. (2004) 'Geography of Droughts and Food Problems in Russia (1900–2000)' in *Report of the International Project on Global Environmental Change and Its Threat to Food and Water Security in Russia* Kassel, Center for Environmental Systems Research

Gorbachev, M. (2000) *On My Country and the World* New York, Columbia University Press

Gorbachev, M. (2001) Interview for the US Public Broadcasting Service (PBS) at www.pbs.org/wgbh/commandingheights/shared/minitext/int_mikhailgorbachev.html

Greenpeace (28 June 2019) 'The *Chernobyl* mini-series has finished, but the real-life catastrophe never ends' at www.greenpeace.org/international/story/22799/the-chernobyl-mini-series-has-finished-but-the-real-life-catastrophe-never-ends/

Gumbert, H. (2011) 'Cold War Theaters: Cosmonaut Titov at the Berlin Wall' in Andrews, J. & Siddiqi, A. *Into the Cosmos: Space Exploration and Soviet Culture* Pittsburgh, University of Pittsburgh Press

Halliday, F. (1997) 'US Policy in the Cold War' in *Socialist History 11 – The Cold War* London, Pluto Press

Harris, C. (1955) 'Industrial and Agricultural Resources' in Inkeles, A. & Geiger, D. (eds.) (1961) *Soviet Society: A Book of Readings* Boston, Houghton Mifflin

Heppenheimer, T. (2002) *The Space Shuttle Decision 1965–72* Washington, Smithsonian

Herwig, C. (2015) *Soviet Bus Stops* London, Fuel

Hill, A. (2017) *The Red Army and the Second World War* Cambridge, Cambridge University Press

History Is Now magazine (21 August 2017) 'Why did the USSR Collapse? Chernobyl, Gorbachev and Glasnost' at www.historyisnowmagazine.com/blog/2017/8/21/the-reason-the-ussr-collapsed-chernobyl-gorbachev-and-glasnost#.X4CLiYuSnIU=

Hobsbawm, E. (1994) *Age of Extremes: The Short Twentieth Century 1914–1991* London, Michael Joseph

Hosking, G. (1985) *A History of the Soviet Union* London, Fontana

Independent newspaper (28 July 2005) 'How Did Yuri Die? The Mysterious Death of a Space-Age Hero' at www.independent.co.uk/news/world/europe/how-did-yuri-die-the-mysterious-death-of-a-space-age-hero-302054.html

International Atomic Energy Agency (IAEA) (1997) *Chernobyl: Facts* at www.iaea.org/Publications/Booklets/Chernoten/facts.html

International Atomic Energy Agency (IAEA) (2006 edition) *Nuclear Power Reactors in the World* at www-pub.iaea.org/MTCD/publications/PDF/RDS2-26_web.pdf

Jenks, D. (2011) 'The Sincere Deceiver – Yuri Gagarin and the Search for a Higher Truth' in Andrews, J. & Siddiqi, A. *Into the Cosmos: Space Exploration and Soviet Culture* Pittsburgh, University of Pittsburgh Press

Ильинский, В., Кузин, В., Саукке, М. (1977) 'Космонавтика на значках СССР 1957–1975 гг' (*USSR Cosmonautic Pin Badges 1957–75*) Москва, Издательство 'Связь'
Каманин, Н. (1972) Лётчики и Космонавты (*Pilots and Cosmonauts* – Nikolai Kamanin's autobiography) Москва, Издательство Политической Литературы
Kamanin, N. (1995–97) *Hidden Space: Letters and Diaries* at http://militera.lib.ru/db/kamanin_np/index.html
Khrushchev, N. (1956) 'The Crimes of Stalin – Speech to the 20th Congress of the CPSU 25th February 1956' in Inkeles, A. & Geiger, D. (eds.) (1961) *Soviet Society: A Book of Readings* Boston, Houghton Mifflin
Khrushchev, N. (1963) *To Avert War, Our Prime Task: Selected Passages 1956–63* Moscow, Foreign Languages Publishing House
Киселев, А., Ребров, М (1967) уходят в космос корабли (*Ships Go Into Space*) Москва, Военное Издательство Министерства Обороны СССР
Kluckhohn, C. (1955) 'Studies of Russian National Character' in Inkeles, A. & Geiger, D. (eds.) (1961) *Soviet Society: A Book of Readings* Boston, Houghton Mifflin
Koenker, D. (2013) *Club Red: Vacation, Travel and the Soviet Dream* Ithaca, Cornell UP
Kolko, G. (1968) *The Politics of War: The World and United States Foreign Policy 1943–45*
Lane, D. (1985) *State and Politics in the USSR* Oxford, Blackwell
Lawton, A. (1992) *The Red Screen: Politics, Society, Art in Soviet Cinema* London, Routledge
Lee, A. (1961) *The Soviet Air Force* London, Duckworth
Lenin, V. (6–8 March 1918). 'Political Report of the Central Committee – Extraordinary Seventh Congress of the R.C.P.(B.)' in *Lenin's Collected Works. 27.* Moscow, Progress Publishers
Lewin, M. (2005) *The Soviet Century* London, Verso
Lewis, C. (2011) 'From the Kitchen into Orbit – the Convergence of Human Spaceflight and Khrushchev's Nascent Consumerism' in Andrews, J. & Siddiqi, A. *Into the Cosmos: Space Exploration and Soviet Culture* Pittsburgh, University of Pittsburgh Press
Lothian, A. (1993) *Valentina: First Woman in Space* Durham, the Pentlands Press
Marples, David R. (May–June 1996). 'The Decade of Despair' in *The Bulletin of the Atomic Scientists.* 52 (3): 20–31.
Maurer, E., Richers, J., Ruthers, M. & Scheide, C. (eds.) (2011) *Soviet Space Culture: Cosmic Enthusiasm in Socialist Societies* Basingstoke, Palgrave Macmillan
Medvedev, R. & Medvedev, Z. (1976) *Khrushchev: The Years in Power* London, Oxford University Press
MacRae, D. (1951) 'The Appeal of Communist Ideology' in Inkeles, A. & Geiger, D. (eds.) (1961) *Soviet Society: A Book of Readings* Boston, Houghton Mifflin
Мил.А, (1984) Знаете, Каким он Парнем Был (*You Know What Kind of Guy He Was*) Москва, Радио и связь
Natural History Museum (2019) *What Is Space Junk and Why Is It a Problem?* at www.nhm.ac.uk/discover/what-is-space-junk-and-why-is-it-a-problem.html
Novosti Press Agency (Moscow) English language booklets: *Man in Space* Moscow (1965), *USSR Probes Space* (1968), *Transfer in Orbit* (1969), *Soviet Cosmonauts* (c. 1970), *Vostok-Soyuz-Salyut* (c. 1971), Glushko, V. *Development of Rocketry and Space Technology in the USSR* (1973), *Baikonur: The World's First Cosmodrome* (1975)
Pankhurst, J. (1988) 'The Sacred and the Secular in the USSR' in Sacks, M. & Pankhurst, J. (eds.) *Understanding Soviet Society* Boston, Allen & Unwin
Pauwels, J. (2002) *The Myth of the Good War: America in the Second World War* Toronto, Lorimer
Riabchikov, E. (1971) *Russians in Space* New York, Doubleday
Ricón, José Luis, (2016) *The Soviet Union: Military Spending*, Nintil, available at https://nintil.com/the-soviet-union-military-spending/
RT (25 May 2009) *BAM: Soviet Construction Project of the Century* at http://rt.tv/Top_News/2009-05-25/BAM__Soviet_construction_project_of_a_century.html

RT World News (30 March 2011) *Gagarin's Undelivered Death Note Published* at www.rt.com/news/note-first-space-life/
Ryabkova. E. (2017) 'ЖЕНЩИНЫ И ЖЕНСКИЙ БЫТ В СССР 1950–1960-Х ГГ. В СОВЕТСКОЙ И СОВРЕМЕННОЙ РОССИЙСКОЙ ИСТОРИОГРАФИИ', 'Women and Women's Life in the USSR 1950–1960 in Soviet and Modern Russian Historiography' in *RUDN Journal of Russian History* 2017 Vol. 16 No. 4, 670–685
Sankova, A. (2020) *Soviet Space Graphics: Cosmic Visions from the USSR* London, Phaidon
Scott, D. & Leonov, A. (2004) *Two Sides of the Moon: Our Story of the Cold War Space Race* London, Simon & Schuster
Siddiqi, A. (2000) *Challenge to Apollo: The Soviet Union and the Space Race 1945–1974* Washington DC, NASA
Siddiqi, A. (2011) 'Cosmic Contradictions – Popular Enthusiasm and Secrecy in the Soviet Space in Programme' in Andrews, J. & Siddiqi, A. *Into the Cosmos: Space Exploration and Soviet Culture* Pittsburgh, University of Pittsburgh Press
Siegelbaum, L. (2011) 'Sputnik Goes to Brussels – The Exhibition of a Soviet Technological Wonder' in Andrews, J. & Siddiqi, A. *Into the Cosmos: Space Exploration and Soviet Culture* Pittsburgh, University of Pittsburgh Press
Sliwinski, M. (1989) 'Afghanistan: Decimation of a People', Orbis, Vol. 33, No. 1, 39–56
Smolders, P. (1973) *Soviets in Space* London, Lutterworth
Smolkin-Rothrock, V. (2011) 'Cosmic Enlightenment – Scientific Atheism and the Soviet Conquest of Space' in Andrews, J. & Siddiqi, A. *Into the Cosmos: Space Exploration and Soviet Culture* Pittsburgh, University of Pittsburgh Press
Soviet Union magazine issues: September and November 1963, July 1954 and No. 136 for 1961
Straubel & the editors of *Air Force* magazine (1959) *Space Weapons* London, Thames & Hudson
Strauss, E. (1969) *Soviet Agriculture in Perspective: A Study of its Successes and Failures* London, George Allen & Unwin
Sykes, G., Mercer, H. & Woolf, J. (1985) *Deadly Persuasion: Teaching the Cold War, a Study of School History Textbooks* London, Teaching the Cold War Study Group
Tereshkova, V. (2015) *The First Lady of Space: In Her Own Words* USA, SpaceHistory101.com
Turbett, C. (2020) *Red Star at War: Victory at all Costs* Yorkshire, Pen & Sword
Turbett, C. (2021) *The Anglo-Soviet Alliance: Comrades and Allies During WW2* Yorkshire, Pen & Sword
US Government (1990) *Soviet Military Power* Washington, Department of Defense
Various (1968) 'Космонавтика – Маленькая энциклопедия' (*Little Encyclopaedia of Cosmonautics*) издательство 'Советская энциклопедия' Москва
Various (1958) *Soviet Sputniks* London, Soviet News Booklet No. 25
Various (1961) *Soviet Man in Space* London, Soviet Booklet No. 78
Various Soviet writers and poets (1977) *The Great Baikal–Amur Railway* Moscow, Progress Publishers
Various (1981) *Understanding Soviet Naval Developments* Washington, Department of the Navy
Воротникова, В.И. (1985) навстречу времени. От Байкала до Амура (*In the Way of Time: From Baikal to Amur*) Москва, Советская Россия
Ward, C. (2009) *Brezhnev's Folly: The Building of BAM and Late Soviet Socialism* Pittsburgh, University of Pittsburgh Press
Евтушенко, Евгений «Просека» (Yevgeny Yevtushenko poem: *Forest Glade*) at https://pitzmann.ru/evtushenko-proseka.htm
Yurchak, A. (2005) *Everything Was Forever until It Was No More: The Last Soviet Generation* New Jersey, Princeton University Press
Зверев, Ю., Оксюта, Г. (1972) Юрий Гагарин На Земле Саратовской (*Yuri Gagarin in the Land of Saratov*) Саратов, Приволжское книжное издательство

Index

A-4 (rocket) 22, 30, 32, 33, 34
Academy of Sciences (Soviet) 22, 95, 110
Alexievich, Svetlana 182, 186, 193
Apollo (US space programme) 31, 81, 122, 134, 139, 141, 142, 148, 149, 199, 205
atheism 18, 120–123, 205
Afghanistan 52, 53, 58, 166, 175, 179–182, 205
Andropov, Yuri 175, 176, 178

Baikal–Amur Railway (BAM) x, xvi, 150, 151, 162–173, 194, 205, 206
Baikonur 61, 62, 82, 101, 104, 110, 137, 139, 141, 189, 190, 204
Belka 71, 72, 79
Belyayev, Pavel 76, 77, 136
Berlin xiv, 45, 46, 54, 56, 107, 188, 194, 201, 203
Bondarenko, Valentin 76, 80–81, 194
Brezhnev Doctrine 180, 187
Brezhnev, Leonid 3, 11, 18, 20, 137, 143, 151, 153, 154, 155, 162, 163, 167, 171, 174, 176, 177, 178, 180, 187, 194, 201, 206
Brussels World Fair 108, 205
Burchett, Wilfrid xv, 73, 115, 201
Bykovsky, Valery 76, 77, 97, 98, 119, 140

Captain America (comic book hero) 17, 50
Chelomey, Vladimir 48, 139
Chernenko, Konstantin 176
Chernobyl 175, 182–186, 201, 202, 203
Chertok, Boris xv, 29, 30, 34, 140, 201
CIA x, 46, 54, 62, 143, 148, 175, 176, 202
Cold War xiv, xviii, 8, 9, 32, 36–58, 59, 60, 61, 108, 148, 172, 175, 178, 180, 184, 192, 201, 203, 205
CPSU x, xvii, 6, 7, 19, 50, 55, 59, 94, 153, 162, 189, 194, 196, 198, 202, 203

CPSU 20th Congress 59, 203
CPSU 22nd Congress xvi, 5, 7, 19, 50, 202
Cuba 42, 43, 47, 58, 115, 116, 166, 197

dogs (in space) 66–72, 79, 110
DOSAAF 10, 20, 80, 88, 94, 96
Dora (Nazi concentration camp) 30, 31

Eisenhower, Dwight 35, 45, 66, 182, 195, 197

Gagarin, First in Space (movie) 83
Gagarin, Yuri xii, 19, 61, 73, 76–78, 81–96, 98, 101, 102, 105, 107, 111–116, 118, 119, 122–125, 128, 132–134, 137, 139, 141–146, 158, 191–193, 195–197, 199, 200–206
Gagarin, Valentina 88–90, 114, 116, 144, 146, 195
Gavin, James 48, 49, 202
Germany xiv, 11, 22, 29, 32, 33, 37, 38, 53, 186, 194, 196, 200
GIRD x, 24, 25, 29, 82, 196, 197
Glasnost 11, 27, 177, 186, 203
Glenn, John 93, 97
Glushko, Valentin 21, 23, 26, 27–29, 33, 34, 110, 142, 196, 204
Gorbachev, Mikhail 13, 51, 53, 54, 56, 57, 174, 177–179, 186, 187, 189, 196, 202, 203
Gulag 26, 138, 201
Gzhatsk 87, 89, 143, 195

HSU (Hero of the Soviet Union) x, 18, 19, 74, 141

Iaquinto, Margaret 190

Jupiter (US rocket) 35, 60, 70

Index

Kaliningrad (Moscow) 34, 61
Kamanin, Nikolai xv, 74–76, 79, 81, 83, 89, 90, 94, 98, 99, 112, 114, 137, 140, 141–143, 196, 197, 203
Kapustin Yar 61
Kartauzov, Leonid 155, 156
Katyusha (rocket) 24, 28, 29
Kaverin, Veniamin 16
Kazakhstan 83, 92, 153, 155, 156, 158, 162, 189, 190, 191, 194, 200
KGB x, 11, 143
Kharkov Planetarium 80, 121–123
Kennedy, John F. 47, 72, 93, 132–134, 141, 197
Kolyma 25–27, 138, 197
Komarov, Vladimir 76, 77, 112, 135, 139, 140, 141, 144, 146
Komsomol x, 19, 20, 80, 88, 95, 96, 154, 157, 158, 163, 164, 166, 167, 169, 170, 173, 193, 196, 199
Korean Airlines (flight 007) 175, 178,
Korolev, Sergei 22–29, 32, 34, 48, 60, 61, 63, 64, 67, 69, 73, 74, 76, 77, 81–83, 91, 92, 94, 97, 101, 104, 110, 111, 113, 116, 128, 133, 134, 138, 139, 142, 143, 144, 146, 191, 196, 197, 202
Kostikov, Andrei 24
Khrushchev, Nikita xii, xvi, xvii, 6, 7, 35, 36, 45, 47, 48, 51, 59, 60, 64, 68, 79, 83, 91, 94, 95, 98, 100, 104, 107, 125, 133, 137, 148, 150–153, 159, 161, 162, 175, 176, 194, 197, 198, 203, 204
Krikalev, Sergei 189, 190,

Laika 68–70
Lenin, Vladimir 37, 109, 139, 150, 192, 204
Leonov, Alexei xv, 61, 75, 77, 78, 101, 118, 129, 136, 141, 143, 148, 193, 198, 205
Lollobrigida, Gina 89, 90
Lunar Programme 134–142

MAD (mutually assured destruction) x, 42, 43, 51, 52
Mars 147
Mashenka (movie) xi
Mir (space station) 189, 190, 191,
Mishin, Vasily 141
Mittelwerk 30, 32

Moon xiii, xiv, 28, 31, 36, 49, 64, 69, 70, 97, 104, 112, 121, 122, 132–142, 146, 149, 198, 199, 201, 205
Moscow Does Not Believe in Tears (movie) 10, 12–14

N-1 (rocket) 134, 139, 141, 142
NASA 10, 15, 65, 66, 93, 133, 141, 147–149, 195, 201, 202, 205
Nazi (Germany) xiii, xiv, 17–19, 21, 22, 26, 29, 30, 31, 34, 35, 37, 38, 53, 111, 194–196, 200
NII-88 (establishment) x, 34
Nikolayev, Andrean 77, 94, 99, 100, 130, 199
NKVD x, 25, 26
Novocherkassk (riot) 161
Novosti Press Agency 23, 55, 63, 66, 69, 72, 101, 106, 110, 136, 204
NPO Energia 189, 190, 191, 196

Office Romance (movie) 14
OKB-1 (establishment) x, 34, 77, 189, 197
Operation Paperclip 30, 32
Operation Osoaviakhim 32
OSOAVIAKhIM x, 94, 95
Outer Space Treaty 49

Peenemünde 30
perestroika 177
Planetariums 80, 121–123
Ponomareva, Valentina 95, 96, 98–100
Powers, Gary 49, 71
Pravda (newspaper) 68, 69, 140, 156

R-2 (rocket) 34
R-5 (rocket) 34, 62
R-7 (rocket) 62–64, 67, 68, 134, 139
RABE (establishment) x, 32
Reagan, Ronald 13, 36, 51, 53, 54, 57, 178, 179, 199
RKK Energia 191
RNII (establishment) x, 24, 25, 29, 32
Roosevelt, Franklin D. 13

Salyut (space station) 146, 147, 204
Saturn (US rocket) 139
science fiction 128
Sharashka (special prison) 26, 29

Sharman, Helen 189
Shepard, Alan 91, 112, 199
Siberia 25, 79, 82, 91, 137, 152, 153, 157, 162, 163, 198
Siddiqi, Asif xv, 48, 94, 97, 98, 114, 201, 203–205
Solaris (movie) 128
Soyuz (rocket) 138, 139–141, 146–149, 189–101, 204
Sputnik (satellite series) 7, 36, 45, 48, 62–72, 81, 84, 86, 88, 106, 109, 124, 125, 128, 174, 195, 201, 205, 206
stagnation (period in USSR) 3, 11, 14, 175, 194
Stalin, Joseph xiii, xiv, 7, 11, 18, 22–25, 32, 34, 37, 38, 40, 41, 43, 45, 59, 60, 67, 120, 124, 151, 152, 161, 162, 167, 177, 198, 201, 203
Star City 77, 81, 89, 104, 116, 143, 146, 195
Star Wars (US space defence programme) 36, 48, 53, 54
Strelka 71, 72, 79
Solovyeva, Irina 95, 96, 98, 99, 100

Tereshkova, Valentina 13, 20, 93–100, 104, 107, 119, 120, 144, 199, 200, 206
The Irony of Fate (movie) 15, 16
Tikhonravov, Mikhail 23, 24, 29, 34, 63, 75
Titov, Gherman 73, 77, 79, 91–94, 97, 98, 101, 107, 111, 112, 115, 122, 123, 142, 200, 201, 203
Truman Doctrine 42
Truman, Harry S. xiv, 41, 42
Tsiolkovsky, Konstantin 21–23, 73, 128, 200
Tupolev, Andrei 24, 26–28, 197

Tushino (airshow) 2, 113–115
Tynda (Siberia) 163, 171
Tyuratam 61, 79

U-2 (aircraft) 47, 50, 61, 62, 71
UK (United Kingdom) 8, 30, 42, 53, 69, 115, 183
USA (United States of America) xii–xiv, 36, 49, 51, 70, 76, 91, 111, 115, 152, 153, 162, 176, 206
Virgin Lands Campaign xvi, 60, 150–162, 170, 194, 201

V-2 (rocket) 22, 30, 33, 34, 70, 194
Venus 146, 147
von Braun, Wernher 22, 30, 31, 32, 34, 35, 60, 194, 195
Voskhod (rocket) 118, 134
Vostok (rocket) 63, 77, 79, 82, 89, 92, 94, 97, 101, 103, 105, 114, 115, 127, 129, 130, 134, 135, 149, 204

White Sun of the Desert (movie) 101
women (in USSR) xiv, 9, 10, 12, 13, 18, 19, 89, 93, 94–98, 101, 116, 119, 156, 157, 164, 166, 168, 170, 174, 179, 199, 202, 205

Yangel, Mikhail 79
Yazdovsky, Vladimir 67
Yeltsin, Boris 189, 191
Yevtushenko, Yevgeny 167–169, 173, 206

Zakharchenko, Vasily 109
Zander, Friedrich 24
Zelenyy (Moscow) 77, 78, 81, 89, 92